D0742769

FLUID ≈ FLOW MEASUREMENT

A practical guide to
accurate
flow measurement

FLUID FLOW MEASUREMENT

A practical guide to accurate flow measurement

E. L. Upp

Published by

Gulf Publishing Company
Houston, London, Paris, Zurich, Tokyo

for information contact

Gulf Publishing Company
Book Division
P.O. Box 2608 Houston, Texas 77252-2608

10 9 8 7 6 5 4 3 2 1

Printed on Acid-Free Paper (∞)

Library of Congress Cataloging-in-Publication Data

Upp, E. L., 1927–
Fluid flow measurement : a practical guide to accurate flow measurement / E. L. Upp.
p. cm.
Includes index.
ISBN 0-88415-017-8
1. Fluid dynamic measurements. 2. Flow meters. I. Title.
TA357.5.M43U66 1993
681′.2—dc20 92-10784
CIP

CONTENTS

DEDICATION

To my wife Ann ...who worked much harder raising our family (including me) than I did during the forty years of my flow measurement career that led to this book.

And deepest appreciation must go to Daniel Industries whose financial support and encouragement made it possible for the book to be created and published.

For over sixty years, the technical personnel of this company have been helping customers solve flow measurement problems. During all this time, it has become more and more apparent that good flow measurement is not a simple commodity to be purchased via a few equipment specifications. That's what this book is all about, and I am pleased to be one of those Daniel employees whose experience and knowledge can continue the tradition that gives great credence to the phrase: Daniel, the experts in flow measurement.

PREFACE

The tendency to make flow measurement a highly theoretical and technical subject overlooks the fact that the *practical* application of meters, metering principles, and metering equipment has the most influence in obtaining quality measurement. And that includes using quality equipment which can continue to function through the years with proper maintenance.

This book is dedicated to condensing and passing along the years of experience that my many friends in the industry and I have accumulated in trying to solve practical flow measurement problems. I cannot begin to name the people that make up my background experience. They include the pioneers in flow measurement, flow measurement design engineers, operating personnel — including top supervisors to the newest testers — academically based engineers, engineers of the manufacturers of flow meter equipment, worldwide practitioners, theorists and people just getting into the business.

And deepest appreciation must go to my friends at Daniel Industries, especially W. A. Griffin. The financial support and encouragement from Daniel made it possible for the book to be created and published.

For over sixty years, the personnel of this company have been helping customers solve flow measurement problems. During this time, it has become more and more apparent that good flow measurement is not a simple commodity to be purchased via a few equipment specifications. That's what this book is all about.

A special thanks is due to Mrs. Patsye Roesler of Daniel Industries, Inc. who typed the document from notes that I could not read and to Jim Anthony who edited and made my Louisiana cajun readable to the English-speaking public.

My personal experience has been that explaining creates the most complete understanding. Standing in front of a class when a student asks you to explain a point clearly separates what you have learned by rote from that which you truly understand. You find out quickly what you really know. Hopefully this book presents to you that which you need to know and understand.

With the large number of early retirements of experienced personnel plus the tendency to make our standards "technically defensible" — but confusing a *practical* guide to successful flow measurement seems to be a useful project.

In those areas covered by the standards, only brief reviews will be made with reference to the documents for additional information. Likewise, detailed theoretical discussions are left to such excellent sources as *Flow Measurement Engineering Handbook* by R. W. Miller[1]. Because of the bulk of such information, data sources are referenced but not reprinted.

While not complete on the total subject, this book should help the reader analyze a flow problem so it may be solved using the other references as needed. I have tried to take as much mystery out of flow measurement as possible by breaking the subject into simple sections and discussing them in everday terms. Each technology has developed its own terminology, and often it seems the purpose is to confuse those not familiar with the terms.

Flow measurement is based on science, but its application depends largely on the art of the practitioner. Too frequently we blindly follow the successful artist simply because "we always did it that way." Industrial experience shows, however, that understanding *why* something is done will generate better flow measurement.

Reference

*1. Miller, Richard W. **Flow Measurement Engineering Handbook.** New York: McGraw-Hill Book Company, 1989.*

1 INTRODUCTION

Chapter Overview

The book's general approach is to look first at basic principles, particularly with respect to differential meters, and the types of fluid flow measurement. Then "theory" is turned into "practice" followed by an overview of fluids and fluid characteristics, both gases and liquids. "Flow" itself is examined next, and finally, comments offered on indiviudal meters and associated equipment.

Emphasis is not so much on individual meter details as on general measurement requirements and the types of meters available to solve the problems.

This first chapter presents some background information, overviews the requisites for "flow" and defines major terms used throughout the book. Chapter 2 introduces various relevant subjects starting with basic principles and fundamental equations. Chapter 3 details the types of fluid measurement — custody transfer and non-custody transfer. Chapter 4 is devoted entirely to listing basic reference standards. Chapter 5 applies theory to the "real world" and describes how various practical considerations make *effective* meter accuracy dependent on much more than simply original manufacturer's specifications and meter calibration. Chapter 6 covers the limitations of obtaining accurate flow measurement because of fluid characteristics. Chapter 7 looks at flow in terms of characteristics required, measurement units involved, and installation requirements for proper meter operation. Chapter 8 reviews meter characteristics, then comments on all major meters. Chapter 9 deals with related readout equipment. Chapter 10 discusses proving systems.

Requisites of Flow Measurement

In this book, "fluids" are common fluids handled in industry in a generic sense, but each fluid of interest must be individually examined to determine if: (a) it is flashing or condensing, (b) has well-defined pressure, volume, temperature (PVT) relationships, (c) has a predictable flow pattern based on Reynolds number (d) is Newtonian, (e) contains no foreign material that will effect the flow meter performance, (i.e. solids in liquids, liquids in gas), and (f) has a measurable analysis that changes slowly, if at all.

The flow should be examined to see if it: (a) has a fairly constant rate or one that does not exceed the variation in flow allowed by the meter system response time, (b) has a non-swirling pattern entering the meter, (c) is not two phase at the meter, (d) is non-pulsating, (e) is in a circular pipe running full, and (f) has provision for removing any trapped air (in liquid) or liquid (in gas) from the meter. Certain meters may have special characteristics that can handle some of these problems, but they must be carefully evaluated to be sure of their usefulness.

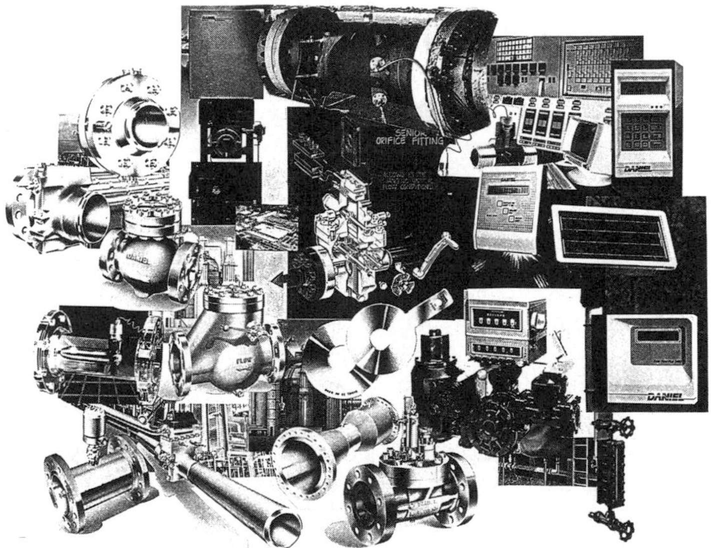

Figure 1-1 *Many different types of meters are available for measuring flow. Proper selection involves a full understanding of all pertinent characteristics relative to a specific measurement job.*

Measurement can usually be accomplished with any one of several meter systems, but for a given job, certain meters have earned acceptance based on their service record. This is an important factor in choosing a meter, and reference to industry standards and users within an industry are important points to review in choosing the best meter.

Background of Flow Measurement

The subjects below form the background for fluid flow measurement and that should be understood before embarking on the task of choosing a flow measurement system. "Fluid", "flow" and "measurement" are defined in generally accepted terms (Webster's New Collegiate Dictionary[7]) as:

Fluid — a. Having particles that easily move and change their relative position without separation of the mass and that easily yields to pressure; b. A substance (as a liquid or a gas) tending to flow or conform to the outline of its container.

Flow — a. To issue or move in a stream; b. To move with a continual change of place among the consistent particles; c. To proceed smoothly and readily; d. To have a smooth uninterrupted continuity.

Measurement — a. The act or process of measuring; b. A figure, extent or amount obtained by measuring.

Combining these into one definition for fluid flow measurement yields:

Fluid Flow Measurement — The measurement of smoothly moving particles, that fill and conform to the piping in an uninterrupted stream, to determine the amount flowing.

Further limitations require that the fluids have a steady state mass flow, are clean, homogenous, Newtonian, and stable with a single-phase non-swirling profile with some limit of Reynolds number (depending on the meter). If any of these criteria are not met, then the measurement tolerances can be affected and in some cases measurement should not be attempted until the exceptions are rectified. These problems cannot be ignored, and expected accuracy will not be achieved until the fluid is properly prepared for measurement. On the other hand, the cost of preparing the flow may sometimes outweigh the value of the flow measurement, and less accuracy should be accepted.

History of Flow Measurement

Flow measurement has evolved over the years in response to demands to measure new products, measure old products under new conditions of flow, and for tightened accuracy requirements.

Figure 1-2 Flow measurement has probably existed in some form since man started handling fluids.

Over 4000 years ago, the Romans measured water flow from their aqueducts to each household to control allocation. The early Chinese measured salt water flow to brine pots used to produce salt used as a seasoning. In each case, control over the process was the prime reason for measurement.

Flow measurement for the purpose of determining billings for total flow developed later.

Well-known names among developers of the head meter are Castelli and Tonicelli in the early 1600's who determined that the rate of flow was equal to the flow velocity times the area, and that discharge through an orifice varies with the square root of the head (pressure drop).

Professor Poleni, in the early 1700's, provided additional work on understanding discharge of an orifice. At about the same time, Bernoulli developed the theorem upon which hydraulic equations of head meters have been based ever since.

In the 1730's, Pitot published a paper on a meter he had developed. Venturi did the same in the late 1790's, as did Herschel in 1887. In London, in the mid 1800's, positive displacement meters began to take form for commercial use. In the early 1900's, the fuel-gas industry started development in the United States (Baltimore GasLight Company).

An early practice in the U.S. was to charge for gas on a per-light basis; this certainly did not reduce any waste, as customers would leave lights on day and night. It is interesting to note that the first positive displacement meters were classified "5-light" and "10-light" meters referencing the size to the number of lights previously counted in a house.

The first of these meters were water-sealed; in the winter, ethanol had to be added to the water to prevent freezing. One of the immediate problems was that not all the ethanol made it into the water baths — and some service personnel found it hard to make it home! In the 1800's, a "dry" type meter was developed which replaced the "wet" meters.

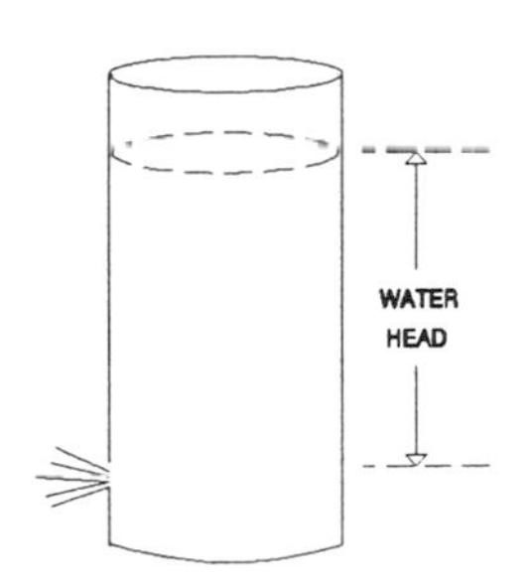

Figure 1-3 *Bernoulli's Theorem for orifice flow from a water pressure head was based on basic laws of physics relating velocity to distance and gravitational force.*

Rotary meters didn't become available until the 1900's. About this same time, Professor Robinson at Ohio State used the pitot to measure gas flows at gas wells. Weymouth calibrated a series of square-edged thin-plate orifices with flange taps. His work was reported in a 1912 paper to the American Society of Mechanical Engineers titled "Measurement of Natural Gas." Similar tests were run on an orifice by Pugh and Cooper.

Also in this time period, Professor Judd at Ohio State conducted tests on concentric, eccentric, and segmental orifice plates. Forerunners of meter companies who also ran tests of their own included Metric Metal Works (later American Meter), the Foxboro Company, and Pittsburgh Equitable (later Rockwell and Equimeter). To study the data and coordinate results, an American Gas Association Committee[1] (1925) began additional testing. This work culminated in AGA Report No.1[2] in 1930 and reported results to-date for the programs being conducted. Work began immediately on Report No. 2[3] which was published in 1935. The first AGA Report No. 3 was published in 1955[4].

The much additional work done since that time is reflected by the latest data in new reports continually being published. The latest, Report No. 3[3] published in 1992, reflects new discharge coefficient data and new installation requirements. Current studies are evaluating the need for further revisions.

Paralleling these efforts is development of meters for use in other areas of flow measurement, meters such as vortex shedding, ultrasonic, magnetic, turbine, and laser.

Flow measurement continues to change as the needs of industry change. No end to such change and improvement is likely as long as mankind uses gas and liquid energy sources requiring flow measurement.

DEFINITION OF TERMS

Accuracy - The ability of a flow measuring system to indicate values closely, approximating the true value of the quantity measured.

Acoustical Tuning - The "organ pipe effect" (reaction of a piping length to a flow-pressure variation to alter the signal). Effects are evaluated based on acoustics.

Algorithm - A step-by-step procedure for solving a problem, usually mathematical.

Ambient Conditions - The conditions (pressure, temperature, humidity, etc.) surrounding a meter, instrument, transducer, etc.

Ambient Pressure, Temperature - The pressure, temperature of the surrounding medium of a flow meter and its recording equipment.

Base Conditions - The conditions of temperature and pressure to which measured volumes are to be corrected. (Same as Reference or Standard Conditions).

Beta - The ratio of the measuring device diameter to the meter run diameter (i.e. orifice bore divided by inlet pipe bore).

Calibration of an instrument or meter - The process or procedure of adjusting an instrument or a meter so that its indication or registration is in close agreement with a referenced standard.

Calorimeter - An apparatus for measuring the heat content of a flowing fluid.

Compressibility - The change in volume per unit of volume of a fluid caused by a unit change in pressure at constant temperature.

Condensing - Reduction to a denser form of fluid (such as steam to water), a change of state from gas to a liquid.

Condensing Point - A measured point in terms of pressure and temperature at which condensation takes place.

Control Signal (Flow) - Information about flow rate that can be transmitted.

Critical Point - That state at which the densities of the gas and liquid phases and all other properties become identical. It is an important correlating parameter for predicting fluid behavior.

Critical Pressure - The pressure at which the critical point occurs.

Critical Temperature - The critical-point temperature above which the fluid cannot exist as a liquid.

Custody Transfer - Flow measurement whose purpose is to arrive at a volume on which payment is made/received as ownership is exchanged.

Dampening - A procedure by which the magnitude of a fluctuating flow or pressure is reduced.

Density - The density of a quantity of homogenous fluid is the ratio of its mass to its volume. The density varies with temperature changes and is, therefore, generally expressed at mass per unit volume at a specified temperature.

Density, Relative (Gas) - The ratio of the specific weight of gas to the specific weight of air at the same conditions of pressure and temperature.

Density, Relative (Liquid) - The relative density of a liquid is the ratio of the substance at a temperature to the density of pure water at a specified base temperature. (This term replaces the term "specific gravity" for a liquid).

Differential Pressure - The drop in pressure across a head device at specified pressure tap locations. It is normally measured in inches or millimeters of water.

Discharge Coefficients - The ratio of the true flow to the theoretical flow. It corrects the theoretical equation for the influence of velocity profile, tap location and the assumption of no energy loss.

Element, Primary - That part of a flow meter under the direct influence of the flow stream.

Element, Secondary - Indicating, recording and transducing elements which measure related variables needed to calculate or correct the flow for variables of the flow equation.

Empirical Tests - Tests based on experience or observed data from experiments.

Energy - The capacity for doing work.

Energy, External - Energy existing in the surroundings of a meter installation. These are normally heat or work energy.

Energy, Flow Work - Energy necessary to make upstream pressure higher than downstream so flow will occur.

Energy, Potential - Energy due to the position or pressure of a fluid.

Energy, Heat - Energy of the temperature of a substance.

Energy, Kinetic - Energy of motion due to fluid velocity.

Energy, Internal - Energy of a fluid due to its temperature and chemical makeup.

Equation of State - The properties of a fluid are represented by equations which relate pressure, temperature and volume. Usefulness depends on the data base from which they were developed and the transport properties of the fluid to which they are applied.

Flashing - Liquids with a sudden increase in temperature and/or a drop in pressure vaporize to a gas flow at the point of change.

Flow Profile - A relationship of velocities in a plane upstream of a meter that defines the condition of the flow into the meter.

Flow Temperature - The average temperature of a flowing stream taken at specified locations in a metering system.

Flow Rate - The volume or mass of flow through a meter per unit time.

Flow Regime - The characteristic flow behavior of a flow process.

Flow, Ideal - Flow that follows theoretical assumptions.

Flow, Totalized - The total flow over a stated period of time such as: per hour, per day, per month.

Flow, Fluctuating - The variation in flow rate that has a frequency lower than the meter-station frequency response.

Flow, Pulsating - The variation in flow rate that has a frequency higher than the meter-station frequency response.

Flow, Slug - Flow with sufficient liquid present so that the liquid collects in low spots and then "kicks over" as a solid slug of liquid. This flow is not accurately measured with current flow meters.

Flow, Layered - Flow that has sufficient liquid present so the gas flows at a velocity above that for liquid flow at the bottom of a line. This flow is not accurately measured with current flow meters.

Flow, Non-Fluctuating - Flow that has no variation in rate except over very long periods of time.

Flow, Non-Swirling - Flow with velocity components moving in straight lines with a swirl angle of less than 2° over the pipe.

Fluid - A substance (liquid or gas) that flows or conforms to its container and moves easily without separation of the mass.

Fluid Flow Measurement - The measurement of smoothly moving particles, that fill and conform to the piping in an uninterrupted stream, to determine the amount flowing.

Fluid Dynamics - Mechanics of the flow forces and their relation to the fluid motion and equilibrium.

Fluids, Separated - Fluids which have been separated into gas and liquids at the temperature of the separating equipment.

Fluids, Dehydrated - Fluids that normally have been separated into gas and liquid with the gas dried to the contract limit by a dehydration unit. (Normally the liquid is not dried, but it may be.)

Frequency Response - The ability of a measuring device to respond to the signal frequency applied to it within a specified limit.

Gas Laws - Relate volume, temperature and pressure of a gas; used to convert volume at one pressure and temperature to another set of conditions, such as flowing conditions to base conditions.

Gaseous Phase - The phase of a substance that occurs at or above the saturated vapor line of a phase diagram. It fills its container and has no level.

Grade, Reagent - Very pure substances that can be considered pure for calculation purposes.

Grade, Commercial - Less-than-pure substances that must meet a composition limit. Although it is normally called by the name of its major component, it is actually a mix.

Head Devices - Meters that use the difference in elevation or pressure between two points in a fluid to calculate flow rate.

Homogeneous Mix - A uniform mixture throughout a flow stream mix, particularly important in sampling a flowing stream for analysis and calculation of fluid characteristics.

Hydrates - Ice-like compounds, formed by water and some hydrocarbons at temperatures that can be above freezing, which block a meter system's flow.

Mass - The property of a body that measures the amount of material it contains and causes it to have weight in a gravitational field.

Mass Meter - Meter that measures mass of a fluid based on a direct or indirect determination of the fluid's weight rate of flow.

Master Meter - A meter whose accuracy has been determined used in series with an operating meter to determine the operating meter's accuracy.

Material Balance - A comparison of the amount of material measured into a process or pipeline compared with the amount of material measured out.

Measurement - The act or process of determining the dimensions, capacity or amount of something.

Meter Proving - The procedure required to determine the relationship between the true volume of fluid measured by a meter and the volume indicated by the meter.

Meter Tube - The upstream and downstream piping of a flow meter installation required to meet minimum requirements of diameter, length, configuration and condition necessary to create a proper flow pattern through the meter.

Meter Inspection - May be as simple as an external visual check, or up to and including a complete internal inspection and calibration of the individual parts against standards.

Meter System - All elements needed to make up a flow meter, including the primary, secondary and related measurements.

Mixture Laws - A fluid's characteristics can be predicted from knowledge of the individual components' characteristics. These mixture laws have limits of accuracy which must be evaluated before applying.

Newtonian Liquids - Liquids that follow Newton's second law which relates force, mass, length, and time. Normal meters measure only Newtonian liquids.

Nozzle - A flow device with an elliptical inlet profile along its centerline and made to a specified standard; usually used for high- velocity flows, resistant to erosion because of its shape.

Phase - A state of matter such as solid, liquid, gas or vapor.

Phase Change - A change from one phase to another (such as a liquid to a gas). Most flow meters cannot measure at this condition.

Physical Constants - The fundamental units adopted as primary measured values for time, mass (quantity of matter), distance, energy, and temperature.

Pipeline Quality - Fluids that meet the quality requirements of contaminants as specified in the exchange contract such as: clean, non-corrosive, single-phase, component limits, etc.

Pressure - The following terms pertain to different categories of pressure.

> Ambient pressure - The pressure of the surrounding medium, such as of the liquid in a pipeline, or of the atmosphere.
>
> Atmospheric pressure - The atmospheric pressure or pressure of one atmosphere. The normal atmosphere (atm) is 101.325 Pa (14.696 psia); the technical atmosphere (at) is 98,066.5 Pa (14.223 psia).

Back Pressure, Turbine Meter - The pressure measured at specified pipe diameters downstream from the turbine flowmeter under operating conditions.

High Vapor Pressure Liquid - A liquid that, at the measurement or proving temperature of the meter, has a vapor pressure equal to or higher than atmospheric pressure (see Low vapor pressure liquid).

Low Vapor Pressure Liquid - A liquid that, at the measurement or proving temperature of the meter, has a vapor pressure less than atmospheric pressure (see High vapor pressure liquid).

Pressure, Gage - Pressure measured relative to atmospheric pressure (atmospheric pressure taken as zero).

Pressure, Impact - Pressure exerted by a moving fluid on a plane perpendicular to its direction of flow. It is measured along the flow axis.

Pressure Loss (drop) - The differential pressure in a flowing liquid stream (which will vary with flow rate) between the inlet and outlet of a meter, flow straightener, valve, strainer, lengths of pipe, etc.

Pressure, Partial - The pressure exerted by a single gaseous component of a mixture of gases.

Pressure, Static - Pressure in a fluid or system that is exerted normal to the surface on which it acts. In a moving fluid, the static pressure is measured at right angles to the direction of flow.

Pressure, Velocity - The component of the moving-fluid pressure due to its velocity; commonly equal to the difference between the impact pressure and the static pressure (see pressure, impact and static).

Reid Vapor Pressure - The vapor pressure of a liquid at 100° F (37.78° C, 311°K) as determined by ASTM D 323-58, Standard Method of Test for Vapor Pressure of Petroleum Products (Reid Method)[6].

Vapor Pressure, (true) - The term applied to the true pressure of a substance to distinguish it from partial pressure, gage pressure, etc. The pressure measured relative to zero absolute pressure (vacuum).

Provers - Devices of known volume used to prove a meter.

Pseudocritical - A gas mixture's compressibility may be estimated by combining the characteristic critical pressures and temperatures of individual components and calculating an estimated critical condition for the mixture.

Pulsation - A periodic, alternate increase and decrease of pressure and/or flow. The effect on a meter depends on the frequency of the pulsation and the frequency response of the meter.

Refined Products - Products that have been processed from raw materials to remove impurities.

Reynolds number - A dimensionless number defined as $(\rho \; d \; v) / \mu$ where ρ is density, d is the diameter of the pipe or device, v is the velocity of the fluids and μ is the viscosity — all in consistent units. Its value is in correlating meter performance from one fluid to another.

Saturation - A state of maximum concentration of a component of a fluid mixture at a given pressure and temperature.

Shrinkage - The amount of loss in apparent volume when two fluids are mixed; caused by the interaction of variable-sized molecules.

Single Phase - One phase (such as liquid without solids or gases present).

Smart Transducers - Transducers with the built-in ability to correct for variations of measured or ambient conditions; an important requirement for most flow meters and measuring devices.

Sour - A fluid that contains corrosive compounds (often sulfur based).

Specific Gravity - (see Density, Relative - gas and liquid)

Specific Weight - The force with which a body at specified conditions is attracted by gravity.

Standard - The following terms pertain to categories of measurement standards:

> Standard - A measuring instrument intended to define, to represent physically or to reproduce the unit of measurement of a quantity (or a multiple or sub-multiple of that unit), in order to transmit it to other measuring instruments by comparison.

International standard - A standard recognized by an international agreement to serve internationally as the basis for fixing the value of all other standards of the given quantity.

National standard - A standard recognized by an official national decision as the basis for fixing the value, in a country, of all other standards of the given quantity. In general, the national standard in a country is also the primary standard.

Primary standard - A standard of a particular measure which has the highest metrological qualities in a given field.

Note:

1) The concept of a primary standard is equally valid for base units and for derived units.

2) The primary standard is never used directly for measurement other than for comparison with duplicate standards or reference standards.

Secondary standard - A standard, the value of which is fixed by direct or indirect comparison with a primary standard or by means of a reference-value standard.

Working standard - A standard which, when calibrated against a reference standard, is intended to verify working measuring instruments of lower accuracy.

Steam, Saturated - The end point of the boiling process. It is the condition where all liquid water has evaporated and the fluid is a gas. Being the end point of the boiling process, its pressure automatically defines its temperature, and conversely its temperature defines its pressure. Saturated steam is unstable; heat loss starts condensation; heat addition superheats; pressure loss superheats; pressure gain starts condensation.

Steam, Superheated - Pressure decrease or heat added to saturated steam will produce superheated steam which acts as a gas and follows general gas laws with increased sensitivity to temperature and pressure measurements.

Steam, Wet (Quality Steam) - A two-phase fluid containing gaseous and liquid water. The quality number defines what part of the mixture is gas; for example, "95% quality steam" indicates that 95% by weight of the mixture is a gas.

Sweet - Fluids containing no corrosive compounds.

Swirling Flow - Flow in which the entire stream has a corkscrew motion as it passes through a pipeline or meter. Most flow meters require swirl to be removed before attempting measurement.

System Balances - (see Material Balance) In a pipeline system this information is reflected in a loss or unaccounted-for report.

Temperature Stratification - At low flow rates, proper mixing does not take place; layers of flow have different temperatures. Proper mixing must be achieved to measure the average fluid temperature.

Uncertainty - A statistical statement of measurement accuracy based on statistically valid information that defines 95% of the data points (twice the standard deviation).

Vapor Phase - The term, used interchangeably with "gas," has various shades of meaning. A vapor is normally a liquid at normal temperature and pressure, but becomes a gas at elevated temperatures. There is also some use of "vapor" to indicate that liquid droplets may be present. In a strict technical sense, however, the terms are interchangeable.

Velocity - Time rate of linear motion in a given direction.

Venturi - A defined head metering device that has a tapered inlet and outlet with a constricted straight middle section.

Viscosity - A fluid's property that measures the shearing stress which depends on flow velocity, which in turn affects the flow pattern to a meter and hence measurement results.

Weight - The force with which a body is attracted by gravity.

Wetted Part - The parts of a meter that are exposed to the flowing fluid.

1. American Gas Association, 1515 Wilson Boulevard, Arlington, Virginia 22209, Publications Department.

2. American Gas Association, Gas Measurement Committee Report No. 1, 1930. 1515 Wilson Boulevard, Arlington, Virginia 22209.

3. American Gas Association, Gas Measurement Committee Report No. 2, 1935. 1515 Wilson Boulevard, Arlington, Virginia 22209.

4. American Gas Association, Gas Measurement Committee Report No. 3, 1955. 1515 Wilson Boulevard, Arlington, Virginia 22209.

5. American Gas Association, Gas Measurement Committee Report No. 3, Part 1 (1990), Parts 2 & 3 (1991), Part 4 (1992). 1515 Wilson Boulevard, Arlington, Virginia 22209.

6. American Society of Testing and Materials D 325-58 Standard Method of Test for Vapor Pressure of Petroleum Products (Reid Method). 1916 Race Street, Philadelphia, Pennsylvania, 19103.

7. Merrian-Webster. Webster's New Collegiate Dictionary. Springfield, Massachusetts: G. & C. Merriam Company.

2

FLOW MEASUREMENT WITH DIFFERENTIAL METERS

All of the following laws should be recognized and met before flow measurement is attempted. Certain physical laws explain what happens in the "real" world. Some of these "laws" explain what happens when fluid flows in a pipeline, and these in turn explain what happens in a flowing stream as it goes through a meter. All variables in the equations must be in consistent units of measure.

Conservation of Mass states that the mass rate is constant. In other words, the amount of fluid moving through a meter is neither added to or taken from as it progresses from Point 1 to Point 2. This is also called the *Law of Continuity*. It can be written in mathematical form as follows:

$$M_1 = M_2 \qquad (1)$$

where: M_1 = mass rate upstream
M_2 = mass rate downstream

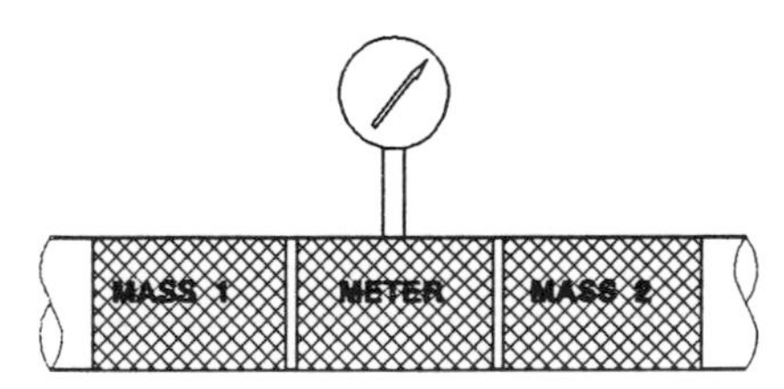

Figure 2-1 The amount of fluid flowing is constant.

Since mass rate equals fluid density times pipe area times fluid velocity, Equation 1 can be rewritten as:

$$\rho_1 A_1 V_1 = \rho_2 A_2 V_2 \quad (2)$$

where: ρ = fluid density at point designated in the pipe
A = pipe area at designated point
V = velocity in the pipe at the designated point
$_{1,\,2}$ = upstream and downstream positions

In terms of volume rate this can be restated as:

$$Q = A * V \quad (3)$$

where: Q = volume per unit time
A, V = as previously defined

Conservation of Energy states that all energy entering a system at Point 1 is also in the system at Point 2, even though one form of energy may be exchanged for another. The total energy in a system combines several types:

1. *Potential Energy* due to the fluid position or pressure.
2. *Flow Work Energy* required for the fluid to flow. The fluid immediately preceding the fluid between Point 1 and Point 2 must be at a slightly higher pressure to exert a force on the volume between 1 and 2.
3. *Kinetic Energy* (energy of motion) due to fluid velocity.
4. *Internal Energy* due to fluid temperature and chemical makeup.
5. *External Energy* is energy exchanged with fluid between Point 1 and Point 2 and the surroundings. These are normally heat and work energies.

The Fluid Friction Law states that energy is required to overcome friction to move fluid from Point 1 to Point 2. For the purpose of calculating flows, **certain assumptions are made about the stability of the system energy under steady flow:** The only energy concerns we have are the potential and kinetic energies (definitions 1. and 3.); the others are either of no importance, do not change between position 1 and position 2, do not occur, or are taken care of by calibration procedures. A generalized statement of this energy balances is given below:

$$KE_1 + PE_1 = KE_2 + PE_2 \quad (4)$$

Kinetic Energy (KE) is energy of motion (velocity). Potential Energy (PE) is energy of position (pressure).

In simple terms, this equation can be rewritten:

$$PE_1 + VE_1 = PE_2 + VE_2 \tag{5}$$

where: PE = pressure Energy
VE = velocity Energy

Equation 5 is the "ideal flow equation" for a restriction in a pipe. In actual applications, however, certain corrections are necessary. The major equation correction is an efficiency factor called "coefficient of discharge."

This factor takes into account the difference between the ideal and the "real world." The ideal equation states that 100% of flow will pass an orifice with a given differential, when in fact, empirical tests indicate that only a fraction of the flow actually passes for a given differential. For example: about 60% with differential between flange taps on an orifice meter, 95% across a nozzle, and 98% across a Venturi. This is because of the device's inefficiency or the loss from inefficiency caused by turbulence at the device where energy of pressure is not all converted to energy of motion. This factor has been determined by industry studies over the years and is reported as "discharge coefficients."

Equations 4 and 5 assume no energy, such as heat, is added or removed from the stream between upstream and the meter itself. This is normally of small concern unless there is significant difference between the flowing and ambient temperatures (i.e., steam measurement), or the measurement of a fluid whose volume is sensitive to very small temperature changes occurring when a fluid is measured near its critical temperature. (Three common examples: ethylene, carbon dioxide gases, and hot water near its boiling point). Also assumed is no temperature change caused by fluid expansion (because of lower pressure in the meter) from the upstream pressure to the meter. The low pressure difference between the two locations normally makes this theoretical consideration insignificant. If there is a change in state (i.e., from liquid to gas or gas to liquid) then this "insignificant temperature change" is no longer insignificant. Furthermore, the volume occupied (assuming no mass hold up) is much greater in the gaseous phase; volume ratios of gas to liquids are as much as several hundred times for some common fluids. Because of these problems, flow measurement of flashing liquids or condensing gases is not attempted.

These equations can be combined and rewritten in simplified forms. Later in this book, the equations will be covered more thoroughly. However, it is

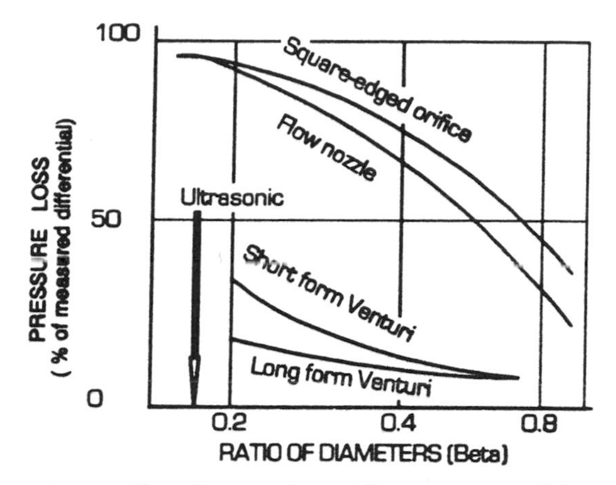

Figure 2-2 Orifice, flow nozzle, and Venturi meters all have pressure loss somewhat less than 100% of full measurement differential; an ultrasonic meter acts like an open section of pipe.

important to recognize the assumptions so that if a metering situation deviates from what has been assumed, a "flag will go up" to indicate that the effect of this variation must be evaluated and treated.

GAS LAWS

Gases are almost always measured at conditions other than standard or base conditions, so they must be converted by calculation using the gas laws to reduce to the desired base conditions .

Boyles Law states that gas volume is inversely proportional to pressure for an ideal gas at constant temperature.

$$V = \frac{Constant}{P} \tag{6}$$

where: V = volume
P = pressure

40 psig
20 psig
10 psig
10 CU FT
20 CU FT
40 CU FT
ALL AT 40°F (500° R)
$P_1 V_1 = P_2 V_2 = P_3 V_3 = K$
10(40) = 20(20) = 40(10) = 400

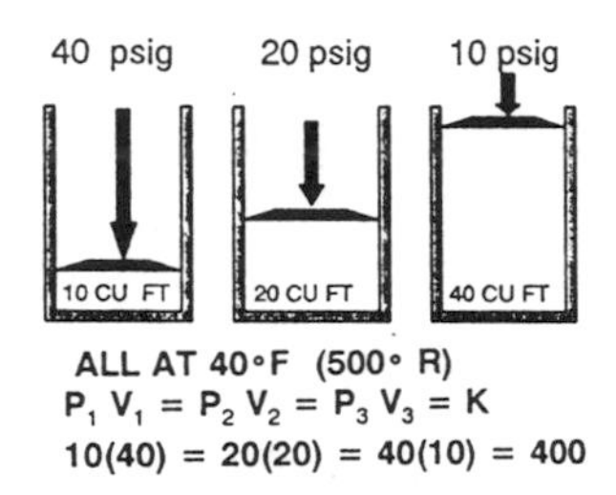

Figure 2-3 *Diagramatic representation of Boyles Law showing that volume is inversely proportional to pressure.*

Charles Law states that gas volume is directly proportional to temperature for an ideal gas at constant pressure.

$$V = Constant(T) \tag{7}$$

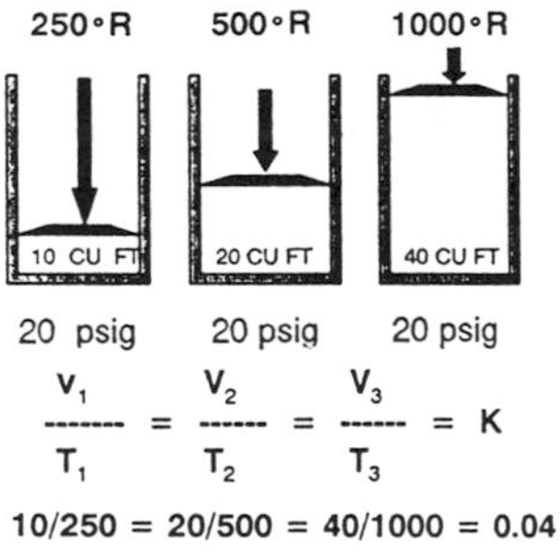

250°R
500°R
1000°R
10 CU FT
20 CU FT
40 CU FT
20 psig
20 psig
20 psig
$\frac{V_1}{T_1} = \frac{V_2}{T_2} = \frac{V_3}{T_3} = K$
10/250 = 20/500 = 40/1000 = 0.04

Figure 2-4 *Diagramatic representation of Charles Law showing that volume is proportional to temperature.*

The Ideal Gas Law combines Boyles and Charles Law; it can be written

$$\frac{V_b}{V_f} = \frac{P_f\, T_b}{P_b\, T_f} \qquad (8)$$

where: $_b$ = base conditions
$_f$ = flowing conditions

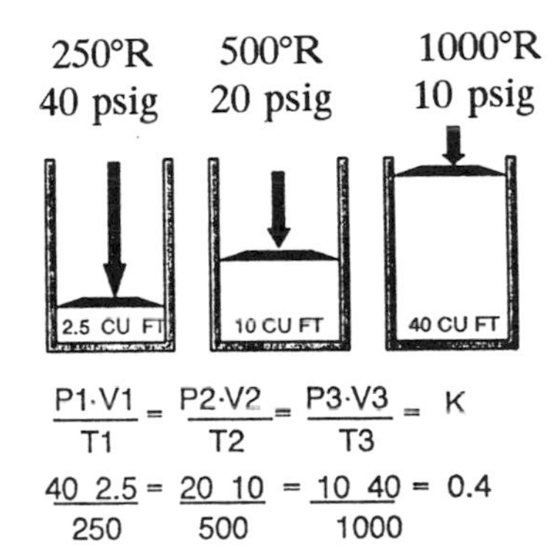

Figure 2-5 *The Ideal Gas Law combines Boyle's and Charles's Laws.*

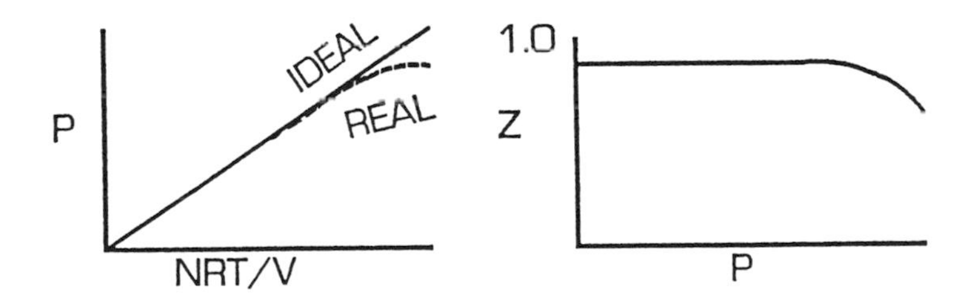

Figure 2-6 *Ideal and actual conditions depart at extremes of pressure and temperature.*

The Real Gas Law (Non-ideal) corrects for the fact that gases do not follow the ideal law at conditions of high pressure and/or low temperature. The ideal gas law equation must be corrected to:

$$\frac{V_b}{V_f} = \frac{P_f\,T_b\,Z_b}{P_b\,T_f\,Z_f} \tag{9}$$

where: Z is the compressibility correction

Empirically derived values for various gases are available in industry standards or are predicted by correlations based on their critical temperatures and critical pressures.

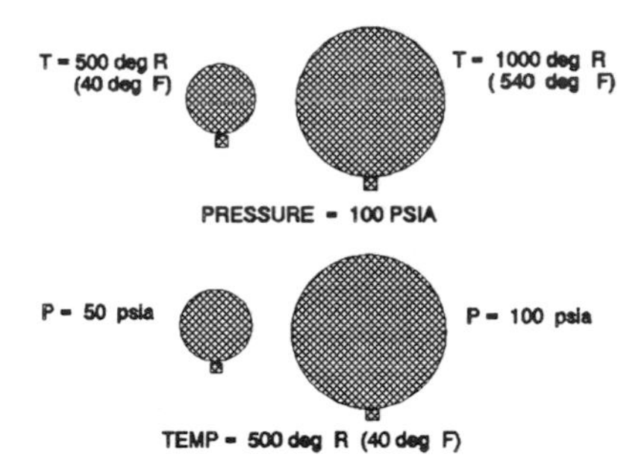

Figure 2-7 Gas-filled balloons also illustrate pressure/temperature/volume relationships.

EXPANSION OF LIQUIDS

Like gases, liquid volumes vary with temperature and pressure. Because liquids have little compressibility with pressure, this effect quite often is ignored unless temperatures approach the critical temperature (within 20%). The effects of temperature are not as large on liquids as on gases. If there is a large difference between the flowing and base temperatures or flowing and base pressures, or if a high degree of accuracy is desired, then corrections should be made.

The *temperature-effect-correction* is based on cubical expansion for each liquid as follows:

$$V_b = V_f [1 + B(T_f - T_b)] \quad \textbf{(10)}$$

where: B = the cubical expansion of the liquid
V = Volume
T = Temperature
$_b$ = base
$_f$ = flowing

The *pressure-effect-correction* is based on compressibility effects for each liquid (this follows API procedures[1]):

$$V_b = V_f \frac{1-(P_b-P_c)F}{1-(P_f-P_e)F} \quad (11)$$

where: P = pressure
$_b$ = base
$_e$ = equilibrium
$_f$ = flowing
F = liquid compressibility correction factor

If $P_b = P_e$ then:

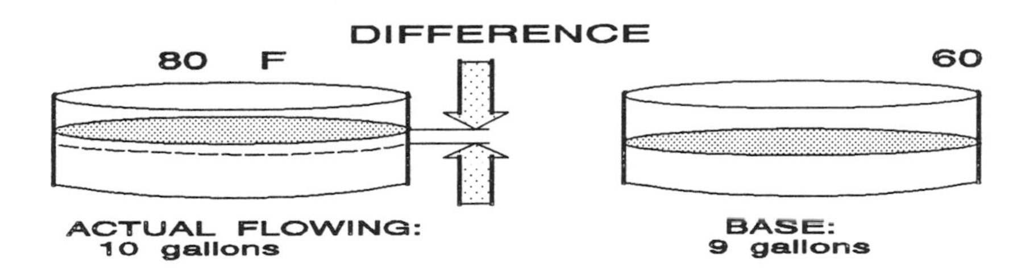

Figure 2-8 Effect of temperature on liquid volume.

FUNDAMENTAL FLOW EQUATION

The fundamental flow equation for differential devices is:

$$q_m = \frac{C\ Y\ \pi\ d^2}{4}\sqrt{2g_c\ \rho_f\ \Delta P} \tag{12}$$

where: q_m = mass flow rate
C = discharge coefficient of flow at the Reynolds Number for the device in question.
Y = expansion factor (usually assumed to be 1.0 for liquids)
π = pi
d = restriction diameter at flowing conditions
g_c = attraction of gravity
ρ_f = density of flowing fluid
ΔP = difference in pressure during flow caused by the restriction. (Note: Discharge coefficient values are available for specific restrictions with specific sets of pressure taps.)

Each standard and each differential device has equations that look different (notations are not the same), but they follow a basic relationship. The fundamental orifice meter mass flow equation in the various standards is:

API 14.3, Sec 3, Part 1[3]

$$q_m = C_d E_v Y(\pi/4)d^2\ \sqrt{2g_c \rho_{tp} \Delta P} \tag{13}$$

AGA-3 1985[1]

$$q_m = KY_1(\pi/4)d^2\ \sqrt{2g_c P_{tb} \Delta P} \tag{14}$$

ISO 5167[4]

$$q_m = CE\epsilon(\pi/4)d^2\sqrt{2\Delta p x \rho_1} \tag{15}$$

ASME[2]

$$q_m = \frac{\pi}{4}C\epsilon d^2\sqrt{\frac{2\Delta p \rho_{F1}}{1-\beta^4}} \tag{16}$$

In these equations:
$C = C_d$
$E_v = E = [1/(1-\beta^{4)}]^{0.5}$
$Y = \epsilon$
$2\Delta P = 2g_c\Delta p$
$K = C_d E_v = CE = c\ [1/(1-\beta^4)]^{.05}$

The equation may be simplified to more easily measured variables (such as differential in inches of water rather than pounds per square inch), and some constants (such as $\pi/4$ and $2g_c$) may be reduced to numbers. If mixed units are used, corrections for these may also be included in the number. Details of this equation for each unit will be covered later in the appropriate section.

The "Volume Flow Rate" may be derived from

$$q_v = \frac{q_m}{\rho_b} \tag{17}$$

where: q_b = volume flow rate at base conditions
q_m = mass flow rate
ρ_b = density of fluid at base conditions

Equations for non-head type volume meters are simpler since they basically reduce the volume determined by the meter at flowing conditions to base conditions as covered in the section on basic laws.

$$q_f F_b = q_b \tag{18}$$

where: q_f = volume at flowing conditions
F_b = factor to reduce from flowing to base (corrects for effects of compressibility, pressure and temperature)
q_b = volume at base conditions

If the meter measures mass, then there is no reduction to base required, and

$$q_{mf} = q_{mb} \tag{19}$$

where: q_{mf} = mass at flowing conditions
q_{mb} = mass at base conditions

References

1. American Gas Association Gas Committee Report No. 3 (1985). 1515 Wilson Boulevard, Arlington, Virginia 22209.

2. American Society of Mechanical Engineers, United Engineering Center, 345 East 47th Street, New York, N.Y. 10017.

3. American Petroleum Institute Manual of Petroleum Measurement Standards, Chapter 14. 1220 L St, N.W., Washington DC 20005.

4. International Standards Organization, Case Postal 36, CH1211 Geneva 20, Switzerland.

3
TYPES OF FLUID FLOW MEASUREMENT

Fluid flow measurement is divided into several types, since each type requires specific considerations of such factors as accuracy requirements, cost considerations, and use of information to obtain the required end results.

CUSTODY TRANSFER

When money is to be exchanged, the "best" flow measurement becomes important so that the two parties to the transactions are treated fairly. The desired accuracy limit for flow measurement is 100% correct. However, no measurement is absolutely accurate; it is simply accurate to some limit.

Q = $

For custody transfer, flow measurement equals dollars.

Figure 3-1 The flow meter is a cash register in custody transfer metering.

In custody transfer metering, being constantly aware that flow measurement is dollars changes the perspective on measurement accordingly. Custody transfer measurement converted to dollars is ± zero.

Quantities for custody transfer are treated as absolute. The responsibility for this measurement, then, is to reduce all inaccuracies to a minimum so that a measured quantity can be agreed upon for exchanging custody.

Control measurement may be accepted at ± 2%; operational measurement may require no more than ± 5%, as contrasted with the ± 0% target for custody transfer metering.

Measurement Contract Requirements

As stated, measurement becomes more complex when two parties must agree to the quantity of product measured and agree to pay/receive money based on this quantity. To protect each party's interest, a contract is normally written that specifies all requirements for measurement of the delivered material such as:

* Definitions used in the contract
* Quantity of material
* Point of delivery
* Material properties
* Measurement Station design
* Measurements to be made
* Material quality
* Price
* Billing
* Force Majeure
* Default or Termination
* Term
* Warranty of Title
* Government requirements
* Arbitration
* Miscellaneous

All these items of interest should be settled prior to commencing measurement for custody transfer purposes. A number of these items are typical information of any contract, but it is of value to expand on the ones that impact the measurement equipment and procedures.

Quantity of Material:
This will specify not only the quantity of the material to be measured by the seller, but also any rights the seller may have to quantities of material above or below the agreed upon amount. This requires the responsible measurement personnel to be aware of these values, to see that contract limits are being met and to ensure that the Seller has the capability of meeting them.

Point of Delivery:
The contract sets forth the point of custody transfer. If the measurement point and the point of delivery are not the same, an agreement must be reached between the Buyer and the Seller for responsibilities for the material between the two points.

Material Properties:
Limits are specified for certain basic properties such as composition limits, pressure and temperature, and actions to be taken are stated if the material is outside the limits.

Measurement Station Design:
The ownership and responsibility for design, installation, maintenance and operation of the meter station of both the Buyer and Seller are spelled out. For metering stations covered by standards, specific references to the standards are made. These standards may be governmental requirements, industry practices or individual company guidelines; they are usually combinations, and detail the kind of meters to be used plus related correcting and readout systems. Details of access by both parties to the equipment and the requirements for frequency of testing and/or reports are spelled out. For large dollar-exchange quantities, there may be an allowance for a check station with the same provisions listed as above stating how any discrepancies between the two measurements will be handled.

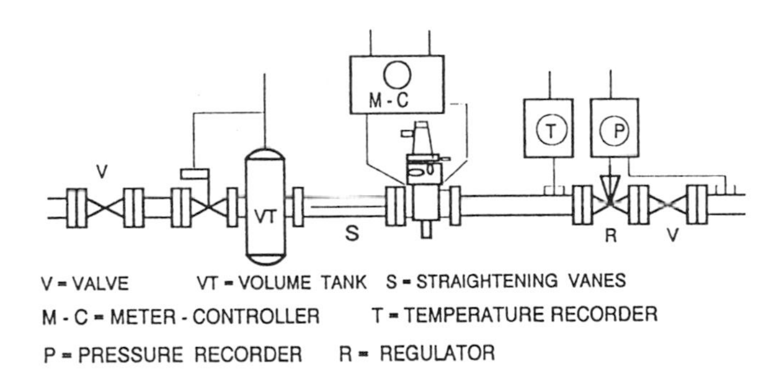

Figure 3-2 Typical meter and regulation station components.

Some provision is made for estimating deliveries during times when the meter is out of service or registering inaccurately, and the procedure for resolving quantities during these periods is included. Some accuracy limits are set; if these are not met as determined by test or check meters, then settlement provisions are implemented.

This accuracy limit may be typically 0.5%, but may be set closer or wider depending on the economics and measurement ability of the meters.

A time period during which a correction can be made is stated if it is not possible to determine the error source and the time of change. Requirements for retaining records and reports are spelled out for both parties. This relates to the specified time period allowed for the quantity measurement to be questioned.

Measurements:

This specifies, in non-confusing terms, the unit of quantity to be delivered. In a volume measurement, base conditions of temperature and pressure are clearly defined. In a weight measurement, only the unit of weight need be specified. For most commercial purposes, the terms "weight" and "mass" are used interchangeably without concern about the effects of the attraction of gravity on the weight being measured. Requirements are specified for all related equipment (beyond the basic meter) and how these secondary measurements will be used to correct basic meter readings. These requirements are particularly important since interested government parties, and the parties to the contract with their plant quantity reports, may record data in different ways. Major confusion can arise if all of these requirements are not spelled out and clearly understood.

Material Quality:

Any natural or manufactured product can have small and varying amounts of foreign material that is not desirable or at least whose quantities must be limited. The quality section defines the rights of the Buyer and the Seller if such limits are exceeded. These specifications may also include separate pricing for mixed streams, so quantities must be delineated for proper payment. If there are too many unwanted contaminants, a price reduction may be allowed rather than curtailing a delivery. These details are spelled out in the quality requirement section.

Billing:

This section sets a deadline for computation of quantity with a provision for correcting errors. It specifies the procedure for billing, the payment period, and penalties for late payment.

Summary of Contract Requirements:

A properly written contract, which protects the interest of the Buyer and the Seller so that fair and equitable billings can be made for an exchange of the quantity of material, is a basic requirement for establishing custody transfer. The ultimate definition of measurement accuracy is achieved when the Sellers sends a bill, the Buyer pays the bill, and both parties are satisfied with the results. All possible misunderstandings and means of their solution should be defined by the contract.

Other Factors in Custody Transfer

Accuracy:

A term used frequently in flow measurement is "accuracy." Accuracy is more abused than correctly used. Unfortunately, it is a sales tool used commercially by both suppliers and users of metering equipment. The supplier with the "best number" wins the bid. Likewise, the user will sometimes require accuracy beyond the capabilities of any meter available. In either case, the accuracy definition serves a purpose for the user or supplier, but has little relevance.

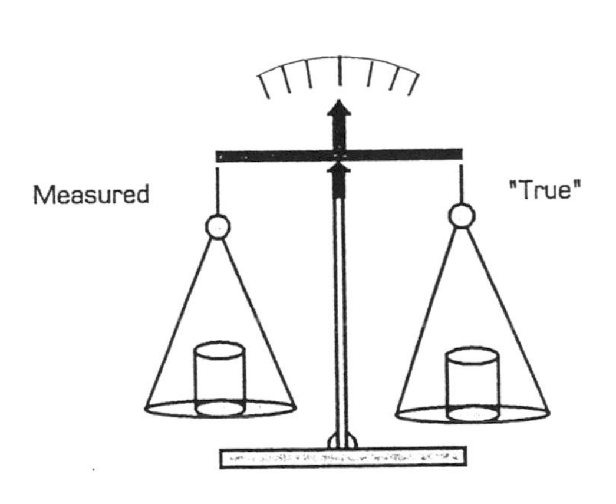

Figure 3-3 "Accuracy" in custody transfer means reducing the difference between measured and "true" flow.

In custody transfer measurement, accuracy is usually defined as the difference between the measured value and the true value expressed as apercentage. The problem with this definition is that the indicated value is read from the meter, but the method of obtaining the true value cannot be specified; therefore, true value is not precisely known. For this reason, it is the subject of many arguments.

The term "uncertainty" with a specified procedure is a statistical statement with at least a comparative meaning when examining various meter capabilities.

Uncertainty:

Performance of the measurement under flowing conditions can be evaluated by making an uncertainty calculation. Many calculation procedures are available in the standards and flow-measurement literature. One is "Measurement Uncertainty for Fluid Flow in Closed Conduits" ANSI/ASME MFC-2M - 1983[2]. The value of this calculation is not in the numerical "absoluteness," but in examining the significance of each variable that impacts the flow calculation and relating these to the flow measurement job in question.

These calculations must consider the particular operating conditions for the specific meter application in order to be most useful in getting the most accurate measurement.

Calculation of the equation's variables is not the total concern for a complete uncertainty determination; allowance must be made for errors from human interpretation, recorders or computers, installation, and fluid characteristics. However, most of these are minimized, provided industry standard requirements are met and properly trained personnel are responsible for operating and maintaining the station. Since these effects cannot be quantified, they are minimized by recognizing their potential existence and properly controlling meter station design, operation, and maintenance. Without proper attention to the total problems, a simple calculation of the equation variables may mislead a user into believing measurement is better or worse than it actually is.

The calculation assumes that the meter has been properly installed, operated and maintained. If maintenance is neglected and the meter has deposits on it that change its flow characteristics, then **the calculation is meaningless** until the meter is cleaned.

Maintenance of Meter Equipment:

Both the supplier and the customer must have confidence that a billing meter is reading the proper delivery volumes. Equipment calibration may change over a period of time, so both parties should take an active part in periodic testing of the meter system. Without tests to reconfirm original accuracies of the metering system, any statement of accuracy is not complete.

The most significant difference in custody transfer metering and in-plant or operation metering is the frequency of equipment maintenance/testing, often weekly. Such tests depend on contractual requirements for type and frequency. They may range from a simple calibration of the readout equip-

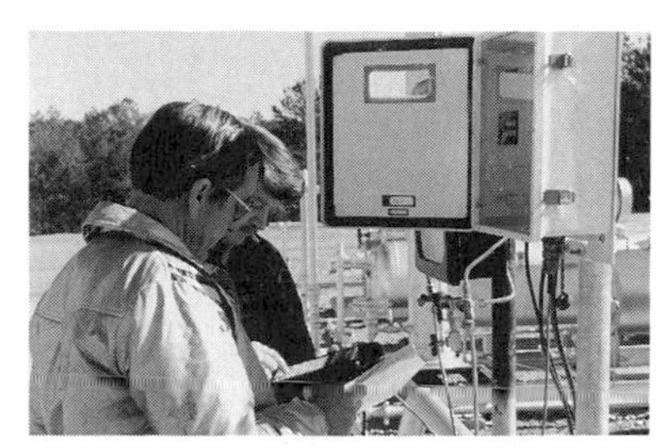

Figure 3-4 Meter system maintenance is vital to accurate flow measurement, and both the buyer and seller should be involved in determining maintenance and proving schedules and procedures.

ment to a complete mechanical inspection of the entire system, or be as thorough as an actual throughput test against some agreed-upon correct volume. In any case, equipment used to test the meter must be approved and agreed upon. Such test devices include certified thermometers for temperature, certified dead-weight testers or test gauges for pressure, certified differential testers for differential meters, certified chromatographs or samples for component analysis, and certified provers for throughput tests. Many models of each are available and can be supplied with accuracy certification. Certification is important to both parties to minimize concern about the equipment.

Another source to build confidence in the calibration equipment and test procedure is the experience of operators of similar metering systems. The test equipment itself should be recertified on a timely basis by the agency or manufacturer that originally certified the equipment.

The first step in testing any meter is visual inspection for any signs of improper operation such as leakage and unstable flow. This includes a review of all of attendant equipment and their indications or recordings. If the station appears to be operating properly, then the individual elements of the station such as the meter and the correctors for pressure, temperature, density, and composition should be individually calibrated with the assumption that if all parts are in calibration, then the system will be in calibration to the limits calculated by the uncertainty equation. This procedure is normally used for standard flow metering.

Proving Meters

If there is a desire to reduce tolerances on this measurement, then an actual throughput test can be run against a "master meter" or a prover system. The master meter should be calibrated and certified to some accuracy limit by a testing facility of a government agency, a private laboratory, a manufacturer, or the user, using agreed upon flow standards. Periodically, the master meter has to be sent back to the testing facility for recertification. Retesting frequency depends upon fluids being tested and treatment of the master meter between tests. The best throughput test is one that can be run directly in series with a "prover".

The prover can come in many forms, but essentially involves a basic volume that has been certified by a government or industry group. Being one step closer to a basic calibration, this is probably the most accurate test of a meter's throughput. Such provers for liquid may be calibrated seraphin cans (for fluids with no vapor pressure at flowing temperature), pressurizedvolume tanks (for fluids with vapor pressure at the flowing temperature), or pipe provers (formerly called mechanical displacement provers as described in API - Manual of Petroleum Measurement Standards)[1]. Such pipe provers may be permanently installed in large-dollar-volume meter stations or may be portable units for testing multiple smaller meter stations.

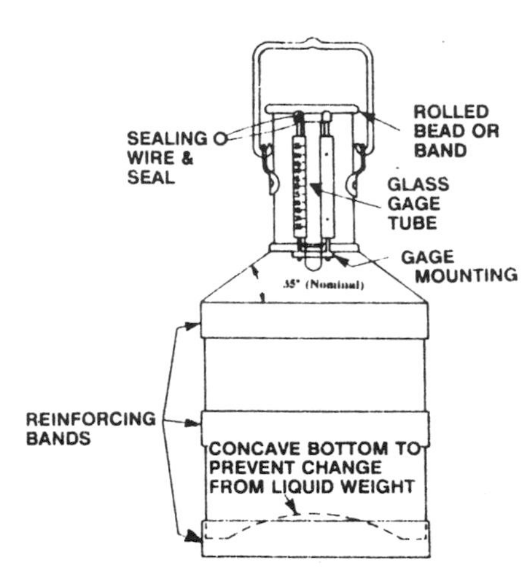

Figure 3-5 Typical seraphin can prover.

Gas provers may be master meters with computer controls so that testing requires little or no calculation or critical flow nozzles (where accurate thermodynamic properties of the gas are available). Critical nozzles require a permanent pressure drop of some 10 to 15% of the upstream static pressure and cannot be run at static pressure below approximately 30 psia.

Figure 3-6 Typical pipe prover for liquid meters.

Great care must be exercised in using such equipment, as detailed by the relevant standards or the manufacturer, to ensure accurate testing. Since these tests are subject to problems, they should be conducted only by qualified technicians. Run correctly, the test ensures the best measurement and provides proof of accuracy.

Proper maintenance and calibration of a billing meter is essential to accurate custody transfer metering. Testing requires participation by both the supplier and the customer. Diagnostics and evaluation with proper test equipment ensure that recorded volumes are correct. Any proving must be documented and signed by both parties so that contract provisions can be implemented on any corrections required.

Properly trained personnel, who understand the importance of the equipment they maintain, are the key to accurate measurement. With proper test procedures, accurate test equipment and a good maintenance procedure, any company should have an acceptable loss or unaccounted-for record.

Operation Considerations:

Choosing the correct meter type is the first step to achieving measurement with a minimum uncertainty. However, the meter's limitation must be recognized to achieve the possible accuracy. Most meters operate within stated limits and should not be used in the extremes of ranges for custody transfer metering.

The first consideration for custody transfer is to minimize flow variations by better control of the flow rate. At times, this may not be possible, and the need for a wide flow range will necessitate further consideration of the flow meter choice. If there is no one meter with the range required to operate in the accurate part of its range, then the use of multiple meters with some type of flow switching control to turn meters in and out of service is required

For example, consider the fuel to a process with three heat exchangers. The range of flow required may be from the exchangers' pilot load to all three exchangers in full load service. This could require a flow range of over 100 to 1. The metering used could be a combination of a positive displacement meter for the low flows and several turbine or orifice meters for the high flows. At one time, this type of complex system could have been a problem, but electronic computers and controllers now make it readily available. Such a system is easily managed, and volumes can be accurately measured with total flow for the whole station reported as a single volume.

In addition to basic meter problems at meter extremes, the secondary equipment that measures pressure, temperature, differential-pressure, density or specific gravity and composition of the flow can also have problems. Normal specifications on these devices are stated as percent of full scale.

Selecting an instrument with the wrong range for the parameters to be measured introduces errors.

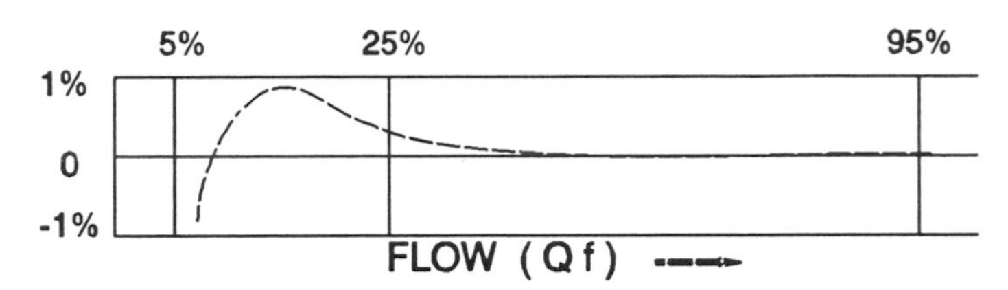

Figure 3-7 For best accuracy, operate this meter above about 25% and below about 95% of full-scale.

The best operating range for metering systems is within the 25% to 95% range of the meter. If operation changes do not over-range the meter, it should be selected to operate near its full-scale reading.

A system properly chosen, designed and installed may still fail to meet expectations if the meter is not designed to operate in its most accurate range.

Custody Transfer Auditing:

When money is exchanged for measured material, the two parties' agreement will include a means of auditing the volumes obtained. Sufficient operation and maintenance records are made available to both parties so that the calculated volumes can be arrived at independently. At least a check of the values used by the other party should be made to see that agreement is reached on the volume.

This procedure is an important aspect of custody transfer metering and is usually completed within 30 to 60 days after a bill is submitted. It will keep both parties involved in the measurement and prevent disagreements about procedures and volumes at some later date. With the data still current, any disagreements can be settled while knowledge of the measurement is fresh in both parties' minds. A complete file of any disagreements must be kept, including resolutions. Records can be reviewed to see if a particular station or particular errors are involved in recurring problems that need to be addressed by an equipment or maintenance upgrade.

Summary of Custody Transfer

Custody transfer measurement begins with a contract between two parties

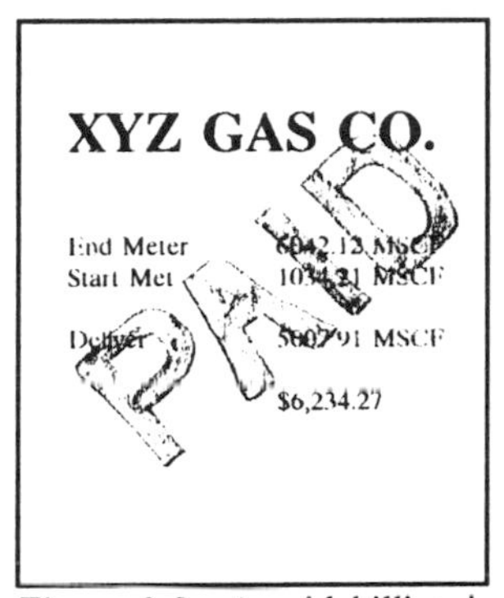

Figure 3-8 A paid billing is the final test of custody transfer flow measurement.

that specifies the data needed to choose a metering system. To get the most accurate measurement required to minimize settlement problems, maintenance and operation of the system must be controlled so that accuracy capabilities of the meter may be realized in service. Information in this chapter should be supplemented by reference to the other chapters of the book for a complete understanding of an individual meter's advantages and limitations. If all precautions are taken, then proof of the station will be when bills are submitted and paid and the custody has been successfully transferred.

NON-CUSTODY TRANSFER MEASUREMENT

Control Signal

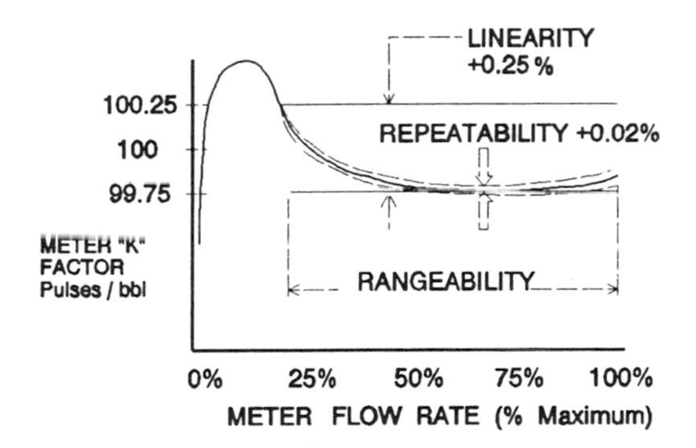

Figure 3-9 Principal terms in defining meter "accuracy."

Of prime importance in any process is the ability to measure flows so that the process can be controlled. The absolute accuracy of such a signal is not as important as the ability to repeat the measurement under the same conditions. It is confusing to speak of not needing accuracy for flow signals as much as needing repeatability, but seldom does a process operate at the original design parameters. There must be a tuning of the process before it comes into balance. Then the job of control signals is to hold the balance and make changes required when the process is varied. An unpredictable meter output can cause control problems, so a control signal must come from a repeatable measurement at a given rate across the range of flow. Because of the sensitivity of some processes to changes, frequency response of a flow control signal is much more critical than for a custody transfer signal. In control measurement, the rate signal represents the process variable of interest whereas, in custody transfer, total flow is required. Therefore, with custody transfer measurement, the ability to integrate volume with time is a significant part of the system; control signals are seldom integrated to totalize volume.

Other Uses

Flow may be measured periodically to check an operation with the assumption that it will then run properly until results indicate otherwise. A good example is heating and cooling distribution in ducts. Once set, it is changed only if there is an indication that the distribution has changed as heating or cooling gets out of control. Similar tests are done for pollution studies. In these "other uses," accuracy may be no better than ± 5 to 10 percent.

Summary of Flow Measurement

Many different capabilities are required to measure flow. Each job should be defined so that expectations of accuracy can be balanced against cost to derive the most cost-effective installation that will do the job required.

References

1. American Petroleum Institute Manual of Petroleum Measurement Standards. API:Washington, DC (Chapters are continually updated).

2. ANSI/ASME MFC-2M Measurement Uncertainty for Fluid Flow in Closed Conduits. New York: Society of Mechanical Engineers, 1988.

4
BASIC REFERENCE STANDARDS

Flow has been measured since the earliest of times. For example, flow measurement was used to control taxing of land owners in the Nile delta. Deposits of rich soil occurred each year during flooding. The amount of land flooded was related to the height of the water level on a calibrated stick whose calibration was under control of the King. At the height of the flood, the stick would be read and taxes for the coming year set on the basis of how much land would be enriched by the flood. When budgets got a little tight, the King simply recalibrated the stick!

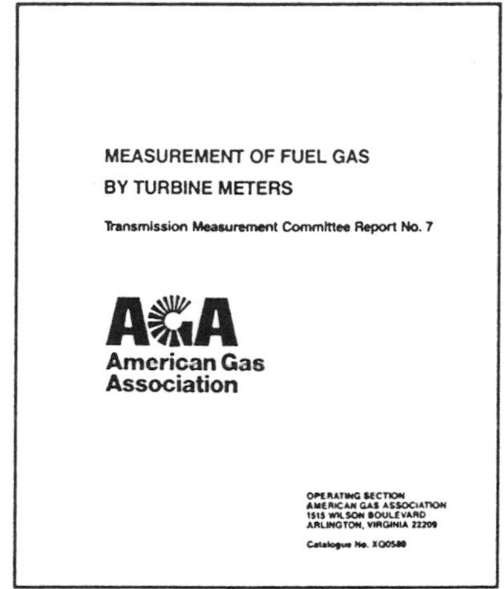
MEASUREMENT OF FUEL GAS
BY TURBINE METERS

Transmission Measurement Committee Report No. 7

AGA
American Gas
Association

OPERATING SECTION
AMERICAN GAS ASSOCIATION
1515 WILSON BOULEVARD
ARLINGTON, VIRGINIA 22209

Catalogue No. XQ0580

Figure 4-1 A typical AGA standard.

The standards for measurement over the years represent continued upgrading of knowledge on older meters and establishment of new standards for newly developed meters. Some of the organizations that have been involved in writing these standards are: American Petroleum Institute (API), American Gas Association (AGA), American Society of Mechanical Engineers (ASME), Gas Processors Association (GPA)[4], Instrument Society of America (ISA), American National Standards Institute (ANSI)[1], American Society of Testing Materials (ASTM)[2], Institute of Petroleum (IP)[5], British Standards Institute (BSI)[3], International Standards Organizations (ISO)[7], and International Organization for Legal Metrology (IOML)[6].

In addition to these standards organizations, much published and widely used data is available from various universities, measurement schools, manfacturers, organizations, books and individuals. Individuals represent a particularly useful source for information on the metering of specialty fluids which may or may not be included in a standard. For new meters, information from the manufacturer must be used; because new meters have only limited use, detailed industry information is scarce.

Once a meter is widely used and has established a track record, standards are written. Specific examples of standards are listed below. (Dates of issue are valid at the time this book was published.) ANSI, like most other organizations involved in flow measurement, requires that standards be reviewed once every five years and either reconfirmed or updated based on the responsible committee's action. Therefore, it is important to check with the organization to ensure that the latest version is ordered and used.

AMERICAN PETROLEUM INSTITUTE (API)

API Ordering Address: American Petroleum Institute, Publications and Distribution, 1220 L. Street, NW, Washington, DC 20005 Phone: (202) 682-8375

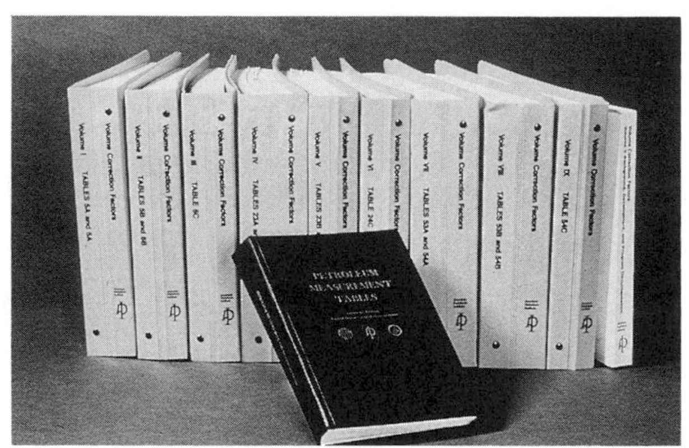

Figure 4-2 *Some of the many volumes in the API "Manual of Petroleum Measurement Standards."*

The API maintains a comprehensive API Manual of Petroleum Measurement Standards. This manual is an ongoing project, as new chapters and revisions of old chapters are released periodically. Listed here are chapters pertinent to flow measurement, specifically as it relates to petroleum products.

Chapter 1, Vocabulary, Reaffirmed August 1987 — This chapter defines and describes the words and terms used throughout the API Measurement Manual. #852-30011

Chapter 4, Proving Systems — This chapter serves as a guide for the design, installation, calibration, and operation of meter proving systems.

Chapter 4.1, Introduction, First Edition, July 1988 — This chapter is a general introduction to the subject of proving, the procedure used to determine a meter factor. #852-30081

Chapter 4.2, Conventional Pipe Provers, First Edition, October 1988 — This chapter outlines the essential elements of unidirectional and bidirectional conventional pipe provers and provides design, installation, and calibration details for the types of pipe provers currently in use. #852-30082

Chapter 4.3, Small Volume Provers, July 1988 — This chapter outlines the essential elements of a small volume prover and provides descriptions of and operating details for the various types of small volume provers that meet acceptable standards of repeatability and accuracy. #852-30083

Chapter 4.4, Tank Provers, October 1988 — This chapter specifies characteristics of tank provers that are in general use and the procedures for their calibration. This publication does not apply to weir-type, vapor-condensing, dual-tank water-displacement, or gas-displacement provers. #852-30084

Chapter 4.5, Master-Meter Provers, October 1988 — This chapter covers the use of both displacement and turbine meters as master meters. #852-30085

Chapter 4.6, Pulse Interpolation, July 1988 — This chapter describes how the double-chronometry method of pulse interpolation, including system operating requirements and equipment testing, is applied to meter proving. #852-30086

Chapter 4.7, Field-Standard Test Measures, October 1988 — This chapter outlines the essential elements of field-standard test measures and provides descriptions and operating details. The volume range of measures in this chapter is 1-1500 gallons. #852-30087

Chapter 5, Metering — This chapter covers the dynamic measurement of liquid hydrocarbons, or metering. It is divided into subchapters as follows.

Chapter 5.1, General Considerations for Measurement by Meters, November 1987 — This chapter is an overall introduction to Chapter 5, Metering. #852-30101

Chapter 5.2, Measurement of Liquid Hydrocarbons by Displacement Meters, November 1987 — This chapter describes methods of obtaining accurate measurements and maximum service life when displacement meters are used to measure liquid hydrocarbons. #852-30102

Chapter 5.3, Measurement of Liquid Hydrocarbons by Turbine Meters, November 1987 — This chapter defines the application criteria for turbine meters and discusses appropriate considerations regarding the liquids to be measured, the installation of a turbine metering systems, and the performance, operation, and maintenance of turbine meters in liquid-hydrocarbon service. #852-30103

Chapter 5.4, Accessory Equipment for Liquid Meters, November 1987 — This chapter describes characteristics of accessory equipment that is generally used with displacement and turbine meters in liquid-hydrocarbon service. #852-30104

Chapter 5.5, Fidelity and Security of Flow Measurement Pulsed-Data Transmission Systems. Reaffirmed August 1987 — This chapter provides a guide to the selection, operation, and maintenance of pulse-data, cabled transmission systems for fluid metering systems to provide the desired level of fidelity and security of transmitted data. #852-30105

Chapter 6, Metering Assemblies — This chapter discusses the design, installation and operation of metering systems for coping with special situations in hydrocarbon measurement.

Chapter 6.1, LACT Systems, Second Edition, May 1991 — This chapter serves as a guide for the design, installation, calibration, and operation of lease automatic custody transfer systems. #852-30121

Chapter 6.2, Loading Rack and Tank Truck Metering Systems, Reaffirmed March 1990 — This chapter guides the selection and installation of loading rack and tank truck metering systems for most gasoline and oil products other than liquefied petroleum gas. #852-30122

Chapter 6.3, Service Station Dispensing Metering Systems, Reaffirmed March 1990 — This chapter covers service station metering systems used for dispensing motor fuel (except liquefied petroleum gas fuels) to road vehicles at relatively low flow and pressure. #852-30123

Chapter 6.4, Metering Systems for Aviation Fueling Facilities, Reaffirmed October 1990 — This chapter is a guide to the selection, installation, performance, and maintenance of metering systems for aviation fuel dispensing systems. #852-30124

Chapter 6.5, Metering Systems for Loading and Unloading Marine Bulk Carriers, Second Edition, May 1991 — This chapter deals with the operation and special arrangement of meters, provers, manifolding, instrumentation, and accessory equipment used for measurement in loading and unloading marine bulk carriers. #852-30125

Chapter 6.6, Pipeline Metering Systems, Second Edition, May 1991 — This chapter provides guidelines for selection of type and size of meters to measure pipeline oil movements, as well as the relative advantages and disadvantages of three meter proving methods. #852-30126

Chapter 6.7, Metering Viscous Hydrocarbons, Second Edition, May 1991 — This publication serves as a guide for the design, installation, operation and proving of meters and their auxiliary equipment used to meter viscous hydrocarbons. #852-30127

Chapter 7, Temperature Determination — This chapter covers the sampling, reading, averaging, and rounding of the temperature of liquid hydrocarbons in both the static and dynamic modes of measurement for volumetric purposes. Portions of Chapter 7 are in preparation. The following chapters and standard now cover the subject of temperature determination.

Chapter 7.2, Dynamic Temperature Determination, June 1985 — This section describes the methods and practices used to obtain flowing temperature using portable electronic thermometers in custody transfer of liquid hydrocarbons. #852-30142

Chapter 7.3, Static Temperature Determination Using Portable Electronic Thermometers, Reaffirmed March 1990 — This section provides a guide to the use of portable electronic thermometers to determine temperatures for custody transfer of liquid hydrocarbons under static conditions. #852-30143

Chapter 8, Sampling — This chapter covers standardized procedures for sampling crude oil or its products. It is divided into subchapters as follows.

Chapter 8.1, Manual Sampling of Petroleum and Petroleum Products, October 1989 — This chapter covers the procedures for obtaining representative samples of shipments of uniform petroleum products, except electrical insulating oils and fluid power hydraulic fluids. It also covers sampling of crude petroleum and nonuniform petroleum products and shipments. It does not cover butane, propane, and gas liquids with a Reid Vapor Pressure (RVP) above 26. #852-30161

Chapter 8.2, Automatic Sampling of Petroleum and Petroleum Products, Reaffirmed August 1987 — This chapter covers automatic procedures for obtaining representative samples of petroleum and nonuniform stocks or shipments, except electrical insulating oil. #852-30162

Chapter 9, Density Determination — This chapter, describing the standard methods and apparatus used to determine specific gravity of crude petroleum products normally handled as liquids, is divided into subchapters as follows:

Chapter 9.1, Hydrometer Test Method for Density, Relative Density (Specific Gravity), or API Gravity of Crude Petroleum and Liquid Petroleum Products, Reaffirmed August 1987 — This chapter describes the methods and practices relating to the determination of the density, relative density, or API gravity of crude petroleum and liquid petroleum products using the hydrometer method (laboratory determination). #852-30181

Chapter 9.2, Pressure Hydrometer Test Method for Density, Relative Density, Reaffirmed August 1987 — This chapter provides a guide for determining the density or relative density (specific gravity) or API gravity of light hydrocarbons, including liquefied petroleum gases, using the pressure hydrometer. #852-30182

Determination of Water and Sediment

Chapter 10, Sediment and Water — This chapter describes methods for determining the amount of sediment and water, either together or separately. Laboratory and field methods are covered as follows:

Chapter 10.1, Determination of Sediment in Crude Oils and Fuel Oils by the Extraction Method, Reaffirmed August 1987 — This publication specifies a method for the determination of sediment in crude petroleum by extraction with toluene. #852-30201

Chapter 10.2, Determination of Water in Crude Oil by Distillation, Reaffirmed August 1987 — This publication specifies a method for the determination of sediment in crude petroleum by distillation. #852-30202

Chapter 10.3, Determination of Water and Sediment in Crude Oil by the Centrifuge Method (Laboratory Procedure) Reaffirmed August 1987 — This publication describes the method of laboratory determination of water and sediment in crude oil by means of the centrifuge procedure. #852-30203

Chapter 10.4, Determination of Sediment and Water in Crude Oil by the Centrifuge Method (Field Procedure), May 1988 — This chapter describes procedures for the determination of water and sediment in crude oils using the field centrifuge procedure. #852-30204

Chapter 10.5, Determination of Water in Petroleum Products and Bituminous Materials by Distillation, March 1990 — This publication describes the test method for determination of water in petroleum products and bituminous materials by distillation. #852-30205

Chapter 10.6, Determination of Water and Sediment in Fuel Oils by the Centrifuge Method (Laboratory Procedure), March 1990 — This publication describes the test method for laboratory determination of water and sediment in fuel oils by centrifuge. #852-30206

Flow Data

Chapter 11, Physical Properties Data — Because of the nature of this material, it is not included in the complete set of measurement standards. Each element of Chapter 11 must be ordered separately. Chapter 11 is the physical data that has direct application to volumetric measurement of liquid hydrocarbons. It is presented in tabular form, in equations relating volume to temperature and pressure, computer subroutines, and magnetic tape.

Chapter 11.1, Volume I, Reaffirmed August 1987 — Table 5A - Generalized Crude Oils and JP-4, Correction of Observed API Gravity to API Gravity at 60^0F

Table 5B - Generalized Products, Correction of Observed API gravity to API Gravity at 60^0F

Table 6-A - Generalized Crude Oils and JP-4, Correction of Volume to 60^0F Against API Gravity at 60^0F #852-27000

Chapter 11.1, Volume II, Reaffirmed August 1987

Table 6B - Generalized Products, Correction of Volume to 60^0F Against API Gravity at 60^0F #852-27015

Chapter 11.1, Volume III, Reaffirmed August 1987 — Table 6C - Volume Correction Factors for Individual and Special Applications, Correction to 60^0F Against Thermal Expansion Coefficients at 60^0 F. #852-37032

Chapter 11.1, Volume IV, Reaffirmed August 1987 — Table 23A - Generalized Crude Oils, Correction of Observed Relative Density to Relative Density at $60/60^0$F.

Table 24A - Generalized Crude Oil, Correction of Volume to 60^0F Against Relative Density 60/60°F #852-27045

Chapter 11.1, Volume V Reaffirmed August 1987 — Table 23B — Generalized Products, Correction of Observed Relative Density to Relative Density at $60/60^0$F #852-27060

Chapter 11.1, Volume VI, Reaffirmed August 1987 — Table 24C Volume Correction Factors for Individual and Special Applications, Volume Correction to 60^0F Against Thermal Expansion Coefficients at 60^0F #852-27085

Chapter 11.1, Volume VII, Reaffirmed August 1987 — Table 53A - Generalized Crude Oils, Correction of Observed Density to Density at 15°C Table 54A - Generalized Crude Oils, Correction of Volume to 15°C against Density at 15°C. #852-27100

Chapter 11.1, Volume VIII, Reaffirmed August 1987 - Table 53B - Generalized Products, Correction of Observed Density to Density at 15°C. Table 54B - Generalized Products, Correction of Volume to 15°C Against Density at 15°C. #852-27115

Chapter 11.1, Volume IX, Reaffirmed August 1987 — Table 54C — Volume Correction Factors for Individual and Special Applications, Volume Correction to 15°C Against Thermal Expansion Coefficients at 15°C #852-27130

Chapter 11.1, Volume X, Reaffirmed August 1987 - Background, development, and computer documentation, including computer subroutines in Fortran IV for all volumes of Chapter 11.1 except Volumes XI/XII, XIII, and XIV. Implementation procedures, including rounding and truncating procedures, are also included. These subroutines are not available through API in magnetic or electronic form. #852-27145

Chapter 11.2.1, Compressibility Factors for Hydrocarbons: 0-90^0 API Gravity Range, Reaffirmed March 1990 — This chapter provides tables to correct hydrocarbon volumes metered under pressure to corresponding volumes at the equilibrium pressure for the metered temperature. It contains compressibility factors related to meter temperature and API gravity (60^{0}F) of metered material. #852-27306

Chapter 11.2.2, Compressibility Factors for Hydrocarbons: 0.350 0.637 Relative Density (60^0 F/60^0 F) and 50^0 F to 140^0 F Metering Temperature, Second Edition, October 1986 — This publication provides tables to correct hydrocarbon volumes metered under pressure to corresponding volumes at equilibrium pressure for the metered temperature. The standard contains compressibility factors related to the meter temperature and relative density (60^0 F/60^{0}F) of the metered material. #852-27307

Chapter 11.2.3, Water Calibration of Volumetric Provers, Reaffirmed March 1990 — This chapter contains volume correction factors in standard units related to prover temperature, and the difference in temperature between the prover and a certified test measure. #852-27310

Chapter 11.3.2.1, Ethylene Density, Reaffirmed August 1987 — This chapter is a computer tape that will produce either a density (pounds/ft^3) or a compressibility factor for vapor phase ethylene over the temperature range from 65^0 to 167^0 and the pressure range from 200 to 2100 psia. The tape is 9-track, 1600 bpi, unlabeled, and is available in either ASCII or EBCDIC format. Format desired must be specified when ordering. #852-25650

Chapter 11.3.3.2, Propylene Compressibility, Reaffirmed August 1987 — This chapter is a computer tape that will produce a table of values applicable to liquid propylene in the following ranges: temperature, 30^0 to 165^{0}F; and saturation pressure to 1600 psia. The tape computes the following two values: density (pounds//ft^3) at flowing temperature and pressure, and ratio of density at flowing conditions to density at 60^{0}F and saturation pressure. The tape is 9-track, 1600 bpi, unlabeled, and is available in either ASCII or EBCDIC format. Format desired must be specified when ordering. #852-25656

Flow Calculation

Chapter 12, Calculation of Petroleum Quantities — This chapter describes standard procedures for calculating net standard volumes, including the application of correction factors and the importance of significant figures. The purpose of standardizing the calculation procedure is to achieve the same result regardless of what person or computer does the calculating.

Chapter 12.2, Calculation of Liquid Petroleum Quantities Measured by Turbine or Displacement Meters, Reaffirmed August 1987 — This publication defines terms used in the calculation of metered petroleum quantities and specifies the equations that allow values of correction factors to be computed. Rules for sequence, rounding, and significant figures are provided, along with tables for computer calculations. #852-30302

Bull. 2509C, Volumetric Shrinkage Resulting from Blending Volatile Hydrocarbons with Crude Oils, Reaffirmed August 1987 — This publication presents data on the subject of volumetric changes resulting from blending volatile hydrocarbons (propane, butane, produced distillates, and natural gasolines) with crude oils. This publication is not included in the current manual. #852-25093

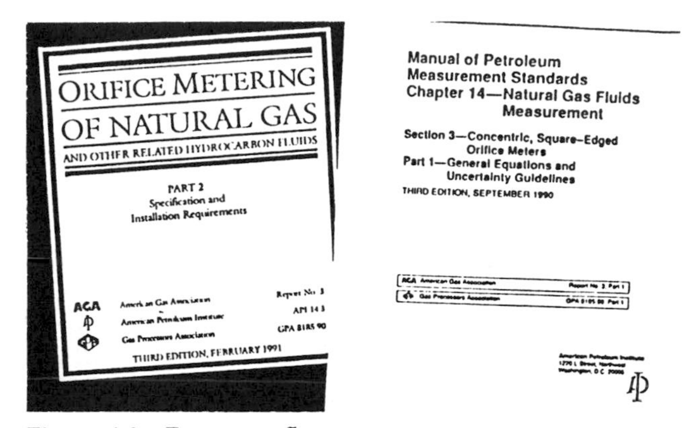

Figure 4-3 For many flow-measurement applications, API and AGA standards are the same.

Chapter 13, Statistical Aspects of Measuring and Sampling — The more accurate petroleum measurement becomes, the more its practitioners stand in need of statistical methods to express residual uncertainties. This chapter covers the application of statistical methods to petroleum measurement and sampling. Chapter 13 is in preparation. The following portion now covers statistical aspects of measuring and sampling, and is included in the manual.

Chapter 13.1, Statistical Concepts and Procedures in Measurement, First Edition, Reaffirmed March 1990 — This chapter is designed to help those who make measurement of bulk oil quantities improve the value of their result statement by making proper estimates of the uncertainty or probable error involved in measurements. #852-30321

Chapter 14, Natural Gas Fluids Measurement This chapter standardizes practices for measuring, sampling, and testing natural gas fluids. Chapter 14 is in preparation. Sections 3, 5, 6, and 8 have been completed and are included in the manual.

Chapter 14.3, Part 1, General Equations and Uncertainty Guidelines, Third Edition, September 1990 (AGA Report #3)(GPA 8185-90) - Part 1 provides basic equations and uncertainty statements for computing flow through orifice meters. In Part 1, the traditional basic orifice factor and Reynolds number factor in the 1985 edition have been repalce with a more fundamental coefficient of discharge that is a function of line size, beta ratio, and pipe

Reynolds number. The upstream expansion factor is not changed from the 1985 edition. The downstream expansion factor has been reanalyzed to include compressiblity. Although each part of the document can be used independently for many applications, users with natural gas applications should review Parts 3 and 4 before implementing Part 1. #852-30350

Chapter 14.3, Part 2, Specifications and Installation Requirements, Third Edition, February 1991. This part; outlines the specification and installation requirements for the measurement of single-phase, homogeneous Newtonian fluids using concentric, square-edged, flange-tapped orifice meters. It provides specifications for the construction and installation of orifice plates, meter tuabes, and associated fittings when designing metering facilities using orifice meters. 32 pages. #852-30351

Chapter 14.5, Calculation of Gross Heating Value, Specific Gravity, and Compressibility of Natural Gas Mixtures from Compositional Analysis, Reaffirmed August 1987 — Outalines procedures to calculate, from compositional analysis, the following properties of natural gas mixtures: heating value, specific gravity, and compressibility factor. #852-30346

Chapter 14.6, Continuous Density Measurement, Second Edition (GPA 8187-87) April 1991 — Formerly titled "Installing and Proving Density Meters," this chapter provides criteria and procedures for designing, installing, operating and calibrating continuous density measurement systems for newtonian fluids in the petroleum, chemical, and natural gas industries. The application of this standard is limited to clean, homogeneous, single phase liquids or supercritical fluids whose flowing density is greater than 0.3 gram per cubic centimeter at operating conditions of 60^0F (15.6^0C) and saturation pressure. Released in 1990. #852-30346

Chapter 14.8, Liquefied Petroleum Gas Measurement, Reaffirmed March 1990 — This chapter describes dynamic and static metering systems used to measure liquefied petroleum gas in the density range of 0.30 to 0.70 grams per cubic centimeter. #852-30348

Chapter 15, Guidelines for Use of the International System of Units (SI) in the Petroleum and Allied Industries, Second Edition, Reaffirmed August 1987 — This publication specifies the API preferred units for quantities involved in petroleum industry measurements and indicates factors for conversion of quantities expressed in customary units to the API-preferred metric units. The quantities that comprise the tables are grouped into convenient categories related to their use. They were chosen to meet the needs of the many and varied aspects of the petroleum industry but also should be useful in similar process industries.

AMERICAN SOCIETY OF MECHANICAL ENGINEERS (ASME)

ASME Ordering Address: American Society of Mechanical Engineers, United Engineering Center, 345 East 47th St., New York, NY 10017 Attn: Publications Department
Phone: (202) 822-1167

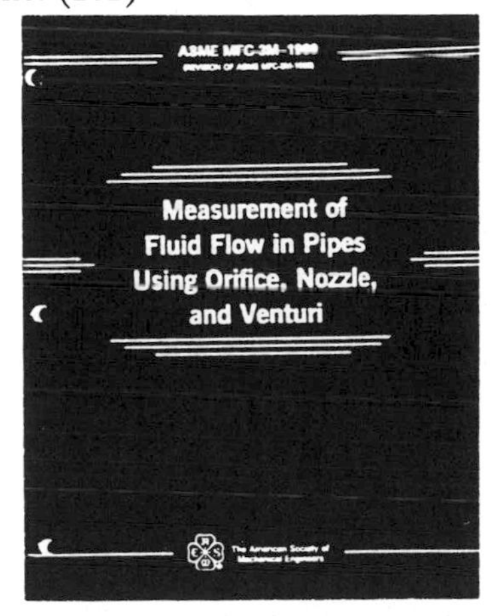

Figure 4-4 Other bodies whose standards impact the flow-measurement industry include ASME, ISA, ISO and others.

MFC-1M-(R1986) Glossary of Terms Used in The Measurement of Fluid Flow in Pipes #J00065

MFC-2M-(R1988) Measurement Uncertainty for Fluid Flow in Closed Conduits #K00112

MFC-3M-1989 Measurement of Fluid Flow in Pipes Using Orifice, Nozzles and Venturi (Not an American National Standard) #K11389

MFC-4M-1986 Measurement of Gas Flow by Turbine Meters #K00118

MFC-5M-1985 Measurement of Liquid Flow in Closed Conduits Using Transit-Time Ultrasonic Flowmeters #K00115

MFC-6M-1987 Measurement of Fluid Flow in Pipes Using Vortex Flow Meters #K00117

MFC-7M-1987 Measurement of Gas Flow by Means of Critical Flow Venturi Nozzles K00119

MFC-8M-1988 Fluid Flow in Closed Conduits-Connections for Pressure Signal Transmissions Between Primary and Secondary Devices #K12188

MFC-9M-1988 Measurement of Liquid Flow in Closed Conduits by Weighing Method #K12588

MFC-10M-1988 Method for Establishing Installation Effects on Flowmeters #K12388

MFC-11M-1989 Measurement of Fluid Flow ISBM 0-7918-2045-9 #K12989

INSTRUMENT SOCIETY OF AMERICA (ISA)

ISA Ordering Address: Instrument Society of America,
PO Box 12277, Research Triangle Park, NC 27709
Phone: 1-800-334-6391

Fundamentals of Flow Measurement, J.P. DeCarlo, 1984 — Provides a basic working knowledge of the methods of flow measurements.

Industrial Flow Measurement, 2nd Edition, D.W. Spitzer, 1990 — Effective flowmeter selection requires a thorough understanding of flowmeter technology plus a practical knowledge of the fluid being measured. This resource reviews important flow measurement concepts to help practicing engineers avoid the costs of misapplication. The text provides explanations, practical considerations, illustrations, and examples of existing flowmeter technology. A rational procedure for flowmeter selection is presented to help decision makers evaluate appropriate criteria.

The ISA Recommended Practice guides cover flow measurement and related instrumentation and are particularly directed to plant operations. A listing of these is available from ISA Publications. In addition to the documents above, ISA offers training courses for self study and also at its facility in North Carolina. Information on these and other services are in the yearly Catalogue of Publications and Training Products.

Flow Measurement, D.W. Spitzer, editor, 1991. This is a part of the Practical Guide Series for Measurement and Control.

ANSI/ISA-RP31.1-1977
Approved 1977

Recommended Practice

Specification, Installation, and Calibration of Turbine Flowmeters

Instrument Society of America

Figure 4-5 Typical Recommended Practice available from the ISA.

Addresses for other sources of measurement standards:

British Standards Institute, 2 Park Street, London W1A 2BS, England Phone (071) 629 9000

International Standards Organization (ISO), Case Postale 36, CH 1211 Geneva 20, Switzerland

International Organization of Legal Metrology, Bureau International de Metrologic Legale, 11 rue Turgot 75009, Paris, France Phone (33(1) 48 481282

5

FROM THEORY TO PRACTICE

A standard is written to be complete. However, between a standard and actual flow measurement, various decisions must be made. It is the application and **use** of the standard that becomes important. For example, ANSI API 2530 (AGA-3)[2] provides a series of drawings reflecting different piping configurations and lengths allowed for different sized orifice plates. (Figure 5-1 below) Applying this section to a design with worst-case parameters (i.e. the longest piping lengths and beta ratio of .75) allows the meter-run piping to be an acceptable length for any other configuration or beta.

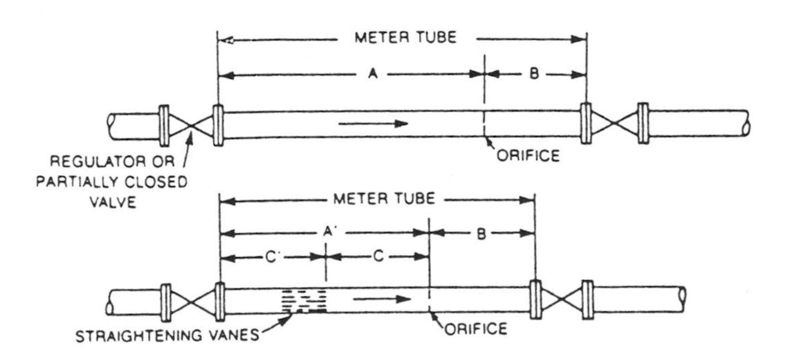

Figure 5-1 *ANSI/API 2530/AGA-3[2], a key orifice-meter document for flow-measurement specialists, provides drawings to help design appropriate meter piping and configurations.*

However, for a specific meter station with space limitations and known maximum volumes, shorter meter tube lengths for lower betas can be acceptable. It would be desirable to have records reflect that this meter tube was designed for a specific set of circumstances so that future users of the tube will be aware of the limitation and not expect it to be a universal length. In process plants where a tube has defined limits that won't change, reduced tube lengths are more common than in the natural gas industry; here the .75 beta figure with the maximum tube length of Figure 5-1 is fairly standard, although tubes length may be even longer in many applications.

Another interpretation of standards is that these lengths were arrived at by empirical tests which indicated that shorter lengths caused the coefficient tolerance to exceed the acceptable limit. Therefore, designs should not be made to minimum limits; they should allow some safety factor by using longer lengths. The interpretation that a design is specified by the standards is not true; the standard simply establishes minimum limits. In other words, each standard is written on the basis of *limits* rather than design specifics.

Those using standards may not fully understand this subtle difference in "what it says" versus "what it means as interpreted by industry application and use." It is worthwhile to seek out these practical interpretations to reach the best design and minimize inaccuracies.

"Ideal" Installations

An "ideal" installation is a worthy objective for a meter station. However, seldom if ever can such an ideal station exist in the real world. Deviation from ideal starts with meter manufacturers defining their meter accuracies based on the best possible conditions of use — and these conditions are not standardized from one manufacturer to another. The result is that a user must insist on a full and complete disclosure of accuracy data derivation to make a legitimate comparison between meters from various makers.

Likewise, a user may not take into account all flow peculiarities (such as dirt or pulsation present), conditions not allowed for in manufacturer data. It is important for the user to know and the manufacturer to be informed of as much of the expected flow information as possible to be able to derive meaningful values for a specific application.

An ideal meter station would be one in which pressure, temperature and flow are stable — both long and short term — changing less than several percent. The fluid should be clean, of non-changing composition with no pulsation, and ample room should exist for the required straight meter tube lengths upstream and downstream. Duplicate instrumentation with automatic switchover to the standby units in case of primary equipment failure should be included. Instrumentation should include automatic transducer testing, and sufficient periodic maintenance should be planned to reconfirm meter calibration. Records should be kept to define any outages or anomalies occurring at the station. And data should be transferred automatically to the billing or other department with all volumes involved accepted by everyone concerned without the need for check meters in series in the line(s).

Seldom does such an ideal station ever exist in real flow measurement. Therefore, allowances must be made for the nonideal characteristics and measurements evaluated accordingly.

NonIdeal Installations

In most cases, requirements for real installations are nonideal. Very few measuring stations have an absolutely pure stream of constant composition to be measured. Truly clean fluids exist only in designer's minds, and some fluid treatment or meter cleaning system must be provided. Most flowing streams have variable flow rates which must be allowed for in the instrumentation. And periodic inspection may or may not be frequent enough for the actual flowing conditions. Flow pressure and temperature normally change with time, if not continually. Space is often limited, so required inlet and outlet lengths have to be compromised. These nonideal conditions can cause considerable loss of accuracy and may well control design considerations.

Station purchase, installation and operation/maintenance costs should, but sometimes do not, reflect what the station will be used for — custody transfer, "company use," line/process control, etc. The standards requirements are not created by considering station errors resulting from noncompliance; they are written so that their requirements are the *minimum* necessary to produce desired measurement accuracy.

Data on installation requirements are part of the background from which each standard is written. If additional knowledge is desired with respect to a standard's application to a particular design, applicable references should be reviewed. Otherwise, the standard takes no position on possible inaccuracy values from each design deviation.

FLUID CHARACTERISTICS DATA

In addition to the flow standards issued by various organizations, related data for fluid characteristics are found in various other references. For example, pressure/volume/temperature (PVT) data are available in the Manual of Petroleum Measurement Standards[3]. In addition, the American Society of Testing Materials (ASTM)[4], American Chemical Society[1], and National Institute of Standards and Technology (NIST)[5] have pertinent data. Also, many correlations have been published by universities as part of advanced-studies programs.

In each case, the data are based on specific parameter limits such as ranges of pressure, temperature and composition. These limits should be known and data used within the limits since extrapolation beyond them may seriously compromise measurement accuracy.

Equations of these correlations for similar products should also be carefully used; results may not agree because of data accuracy limits. Each industry uses "accepted data." From time to time, these data are updated based on additional work. The quality of all such work must be determined before results are used.

Limitations of Accuracy

Of all the questions asked at a gathering of flow measurement personnel, the most frequently asked and the least satisfactorily answered is, "What is the meter accuracy?" It is an unfortunate fact of life that the "one-upmanship" often practiced in both the purchase and sale of flow devices may obscure actual meter performance. To adequately define the problem, the following areas of interest must be considered before any discussion is meaningful:

1. Definition of accuracy
2. Design of equipment and technical limitations
3. Manufacturers' adherence to proper techniques to control making the precision device
4. Installation of the equipment to maintain the manufacturing tolerance
5. Operation of the station in a manner to produce the best measurement accuracy
6. Maintenance required to get long-term measurement performance
7. Proper definition of flowing fluid characteristics
8. The calculation of volumes from the basic equipment.

Definition of Accuracy

(See Chapter 8 for a more complete overview of "accuracy" and other factors involved in meter selection.)

The accepted definition of the term accuracy in measurement of any kind is based on the ratio of the "indicated measurement" to the "true measurement." For flow measurement the ratio is "indicated flow" to "true flow." This seems to be a rather simple problem until an attempt is made to define and demonstrate "true flow." Some definitions of true flow have included:

1) "What the orifice with a recording chart says";
2) "What the tank gauge says";
3) What the government agency says";
4) "What the manufacturer says";
5) "What the lab test says"; or
6) "What I know is right."

All of these, or variations of them, have been used to define true flow, and hence accuracy. The obvious weakness in each is how it allows a wide variety of answers to be obtained. Recently, considerably more "testing to determine various accuracies" has been done by individuals and standards groups. But, even now, not all of results are in agreement.

The flow measurement industry does not have an acceptable statement of how these comparisons of indicated and true measurements should be made. It is becoming more common to use a statement of twice the standard deviation of a statistically valid test-sample population as the uncertainty reported. This in itself is not an absolute statement of what a given meter will do; it simply states how it will do in some 95% of the cases compared to "the most probable value" as determined by test. The test procedure is not specified. The investigator — whether in industry, a manufacturer, or a governmental agency — sets the test conditions. Results may appear "correlated" when fluid is measured once. In industry, however, fluid is normally measured twice: once in and once out; differences then become apparent.

The second area of caution relates to accuracy or uncertainty of a meter system compared to a primary measuring device. The user is interested in overall *system* accuracy (i.e., how good is the number from the system readout), not statements about individual parts — and particularly not a statistical statement based on full scale reading of the parts of a system. Without this understanding of the background of the "accuracy numbers game," it is difficult to evaluate statements about a meter's accuracy made by users and manufacturers.

Most of the numbers that come up in a discussion of flow accuracies are supplied by sources other than the one with the most critical data: the user.

The user, then, should be aware of all pertinent factors involved so that a meaningful estimate of likely measurement accuracy can be made. Properly used flow meters of all types are capable of accuracies that fit in certain categories of proper application. It is the responsibility of those using such meters properly to fit the meters to the actual user needs.

Design and Technical Limitations

All flow meters use fluid-mechanics principles for arriving at flow from the fluid's transport properties. Each of these principles has technical as well as practical limitations.

For example, the orifice meter is one of a category of meter that requires a pressure drop larger than the pressure drop in normal piping for proper measurement. If insufficient pressure drop is available for measurement, then such a head meter cannot be used accurately. This statement seems self-evident; however, users sometimes apply an orifice meter with only a few inches of water differential and still want "accurate flow measurement." Similarly, a meter that senses velocity must sense an accurate average velocity in relation to a known hydraulic area of opening or there is no way to calculate accurate volumes. This means a proper profile must be presented to the meter, and the meter must be kept clean.

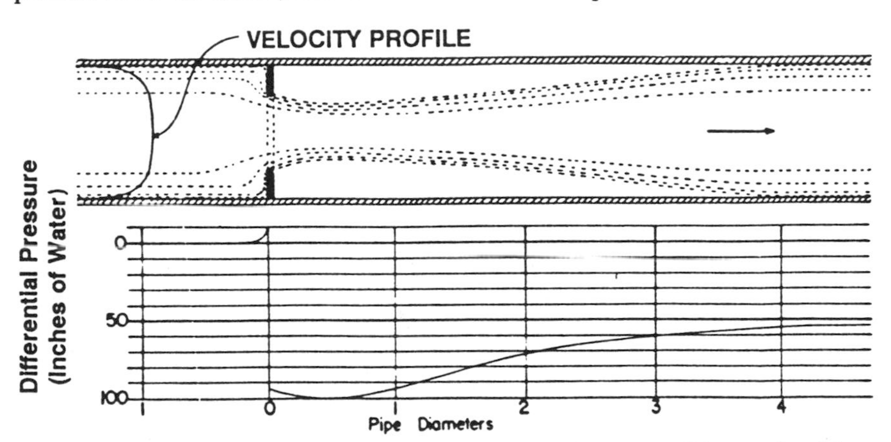

Figure 5-2 Sufficient pressure drop must be created by flowing conditions to be able to derive valid flow measurement.

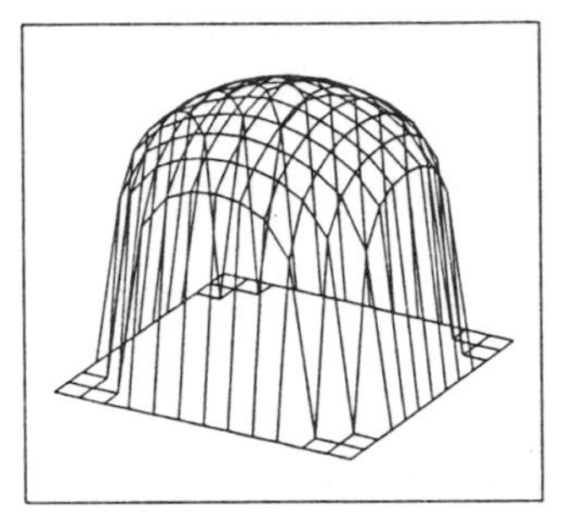
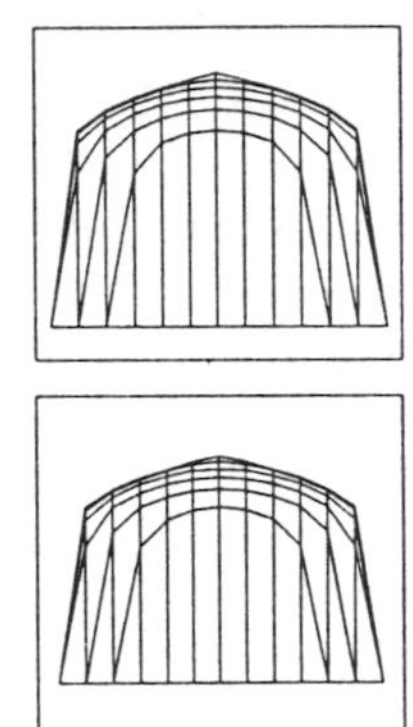

Figure 5-3 Standard fully developed profile.

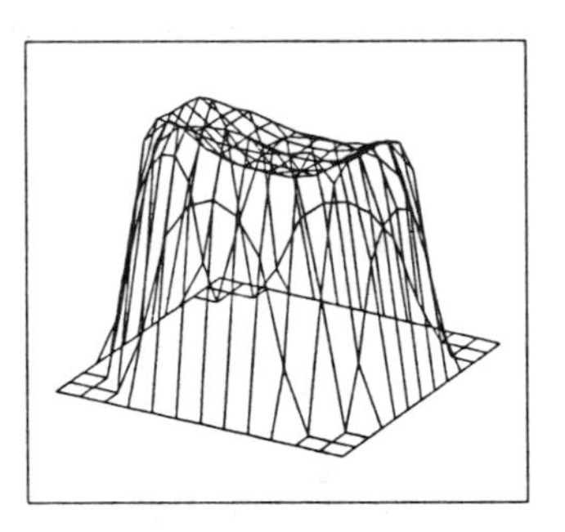
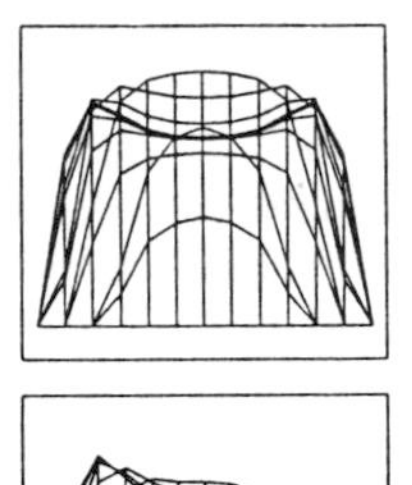

Figure 5-4 Profile following a single elbow.

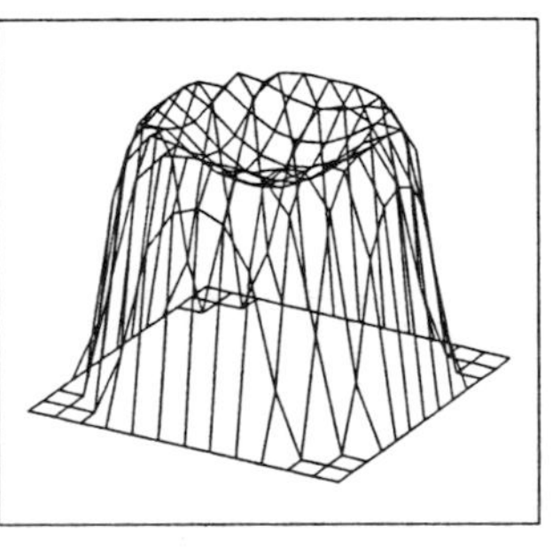
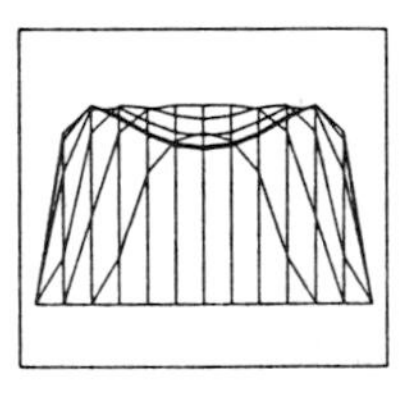
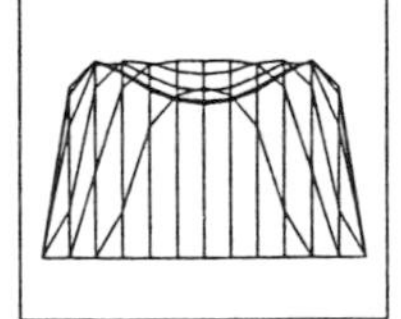

Figure 5-5 Swirling profile — two elbows in different planes.

In general, all flow devices are subject to the following limitations documented in standards or manufacturers' literature. Ignoring these considerations means that any statement of accuracy is meaningless:

- Reynolds number sensitivity
- Nonpulsating flow
- Special piping requirements (flow profile dependency)
- Practical rangeability limits
- Acceptable calibration data
- Single-phase fluids
- Accurate measurement of several variables to arrive at standard volume
- Maintenance requirements with time
- Recording and/or calculating data correctly.

By coordinating the desired measurement with the above information and relevant standards and/or manufacturer's data, intelligent decisions can be made on the possible accuracy that can be expected for a given installation.

Reynolds number relates how one fluid flow behaves in relation to other fluids with the same number. Meters are affected to a large or smaller degree depending on the specific meter's response to the flowing Reynolds number. Reynolds number sensitivity should be checked when considering a meter for a given job.

Pulsating flow presents a problem to most flowmeters. Anytime a designer suspects that pulsation will be present, the meter must be installed with pulsation eliminators between the source of pulsation and the metering device. Work with most commercially used metering devices indicates that virtually no meter is immune from the effects of pulsation.

Piping requirements and flow profile interrelate. Piping adjacent to the meter run can help create a proper, fully developed flow profile. As previously mentioned, lengths specified by various standards should be considered the minimum required, and any additional straight pipe will simply add confidence that the measurement is correct.

Meter manufacturers should — and most do — control all design variables found to affect development of the proper flow profile. However, the care that a manufacturer puts into a meter can be destroyed if proper installation and maintenance procedures are not followed by the user.

Any installation with meter tube lengths less than those required by the standards or manufacturer requirements will not have predictable flow patterns and hence should not be used without an in-place calibration.

The most accurate measurement will occur at the upper range of the meter. Any physical location where measured variables tend to be constant will be a better location than those where wide fluctuations occur. Measurement will be aided by regulating pressure, stabilizing temperature and insuring consistency in flow-stream composition.

The meter tube is defined as the adjacent upstream and downstream piping attached to the meter. Once again, standards or the manufacturer's recommended guidelines should be followed for establishing tolerances in the manufacture of these tubes.

How the primary element is attached to the meter tube is also an important step. For example, fabrication should begin with properly selected pipe or meter-run tubing. Heat from welding can cause distortion at critical points. Unless proper welding techniques are used, a unit that will assure the maximum in measurement accuracy cannot be produced.

The meter and adjacent piping must be properly aligned. If gaskets are used, they should be undercut by approximately one-eighth of an inch to prevent an extrusion of the gasket into the line when bolts are tightened. Seemingly insignificant items of this nature should not be overlooked in providing the most accurate primary measurement device.

***Figure 5-6** Good design calls for long, straight meter-tube lengths for the most accurate flow measurement.*

Recommendations concerning the upstream and downstream piping of the meter tube are covered quite thoroughly in standards and manufacturer's literature. No attempt will be made here to duplicate this coverage other than to emphasize again that meter tube lengths for all measurement conditions will be best obtained by using the extreme conditions as the design standard minimum.

Connecting lines are also important. On head devices with the primary element designed and installed to give an accurate differential at the taps, proper lead lines must also be installed to take advantage inherent in the primary-device accuracy.

Several considerations in the design and installation of these lines for gas measurement must be noted. Taps should come off the top, or at least no more than 45° from the vertical of the gas measuring line. The lines should be large diameter, as short in length as convenient, and with no direction or diameter changes to minimize leaks and pulsation effects. They should be installed with an upward slope away from the line of at least 1 inch per foot of tubing length with the differential device located above the line. This facilitates drainage of any condensable fluids back into the line.

The presence of liquid blockage in sections of these lines can cause errors in the order of the equivalent head of water. All gases at flowing temperatures above ambient with pressures near condensation have this problem. Natural gas saturated with water poses this problem, even though the natural gas itself does not condense. A cold night or a cold rain can cause the entire lead line and instrument to fill with fluid; and then, on warming up again, the fluid will evaporate. During this time the instrument indication is questionable for determining rate or total flow. This is a more critical requirement for the differential device than for the other devices; however, the installation suggestions above will minimize problems for all of the devices.

For liquids, the lines should come off the bottom half of the pipeline, preferably at 45° from the bottom to prevent solids from filling the lines and blocking the differential device. The purpose of the installation is to keep the connecting lines full of liquid even though there occasionally may be gas going down the flowline with the liquid. If a liquid may be heated above its vaporization point by the ambient temperature, then some type of insulation should be installed to control the temperature and maintain the liquid leg.

The flow profile or pattern across the meter is very important for flow measurement accuracy. Two factors control this pattern: (a) piping configuration — including length, roundness, smoothness — and nearest fitting such as elbows, valves, tees, and (b) Reynolds number — which is the guide to the shape, size and stability of the inlet pattern.

Fortunately, most gas is handled at relatively high Reynolds numbers (above 10,000) so that it seldom is a matter of the internal viscous forces becoming a major component of the predominate inertial forces. A high Reynolds number range is one in which the flow pattern easily becomes stabilized provided the piping is properly installed.

Liquids can have a range of Reynolds numbers depending on the liquid viscosity. A liquid that has a higher viscosity than water should be checked to make sure its Reynolds number is higher than required for the particular meter. Some meters are specifically designed to operate at low viscosities if the Reynolds numbers are low.

Practical rangeability limits vary with the meter and measurement conditions. A single meter has a limited range for the accurate determination of flow and this should not be approached at either the high or low extremes. This range can be extended by the use of multiple meters where wide variations in flow are experienced and accurate measurement should be approached in this manner.

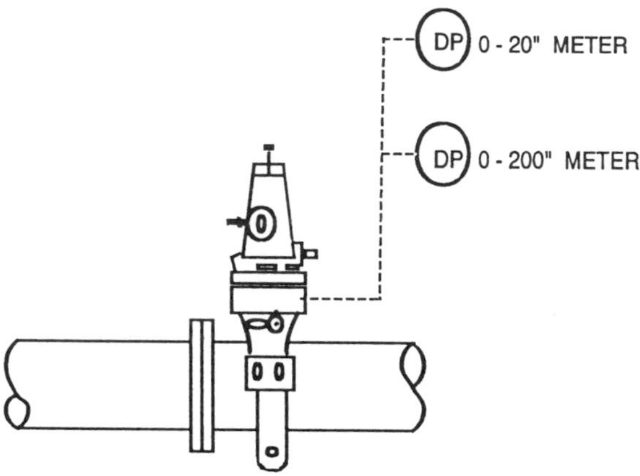

Figure 5-7 *If a meter's rangeability is not sufficient to cover the flow range being measured, multiple transducers can be used or a different type meter evaluated.*

A head meter, as the name implies, requires the sacrifice of some pressure, which is of little significance in some fluid measurement, but can be a severe limitation if pressures are low.

Present standards limit the orifice meter run sizes to 2 inches minimum and 30 inches maximum. Larger sizes not covered in standards are available based on extrapolated data. Turbine meters are available in different limited sizes from various manufacturers, with the largest meter made in the U.S. presently limited to 12 inches in gas applications and 24 inches for liquid applications. European manufacturers make larger gas turbine meters. Rangeability for turbine meters typically runs from about 10- to-1 on liquids, and from this range to more than 100-to-1 on high-pressure gas; for PD meters the rangeability runs from about 20-to-1 to as much as 1000-to-1 on liquids and gases.

Standards calibration data depends upon data collected over time. Some meters are fully covered by industry standards written over many years. Calibration data on them has been tested many times. Newer meters have to go through a period of acceptance and test normally starting with data supplied by the manufacturer and accepted by the user.

Coefficients supplied must be carefully evaluated and checks made frequently on new meters. Once sufficient experience has been obtained and data made available, standards organizations will run their own tests (or accept tests by others) and prepare a standard.

Industry usually accepts either source of calibration data but will be more cautious about using the data with new meters when critical measurement is involved. Standards typically take from five to ten years for completion.

Single-phase flow exists in most practical measurement situations. Flow pattern is affected by the presence of two-phase flow, and density is difficult to determine for a nonhomogenous fluid. There are some approximating methods for measuring two-phase flow with head meters, but the resulting data are not precise. The Coriolis meter can measure two-phase flow over limited ranges.

Measurement of other variables needed to derive accurate standard volume requires attention and understanding equal to that involved with the primary device. The overall accuracy of the flow meter begins with the primary device, but is also dependent on the transducers necessary to obtain the flowing density either directly with a densitometer or indirectly — through measurements of pressure, temperature, compressibility and specific gravity — to convert flowing conditions to base conditions.

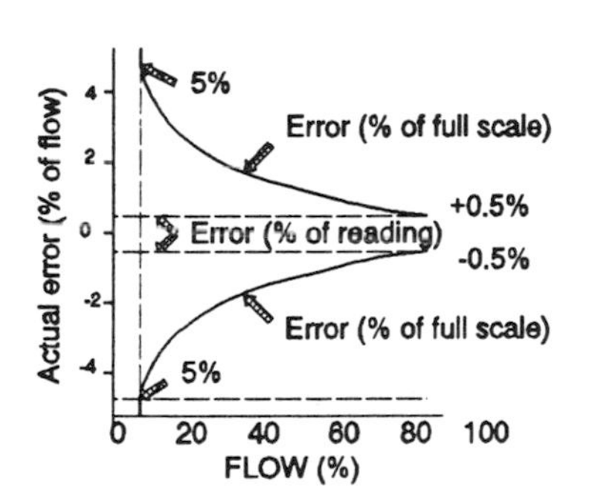

Figure 5-8 Accuracy and other characteristics of transducers are as important as primary meter characteristics for accurate, corrected flow.

Density Accurate calculation of standard volume through a meter requires knowing the density of fluid flowing at the meter and proper interpretation of the measurement through use of appropriate equations to reduce the flow to base conditions. In the past, density was not measured directly but was calculated from a measure of pressure, temperature, specific gravity and compressibility.

Direct measurement of density is now available in one piece of equipment. The density measurement is needed at flow sensitive points such as at the plane of the orifice plate bore or at the turbine wheel in a turbine meter. A densitometer may be installed in a less sensitive location providing correction or control of the variables is made to arrive at the correct density from these remote locations. The end product of the measurement method must be the actual density at the measuring device.

Differential Two of the major sources of error in the application of a head meter come from taking the square root of the differential measurement and from the effects of small errors in the differential at low differentials causing large errors in flow measurement.

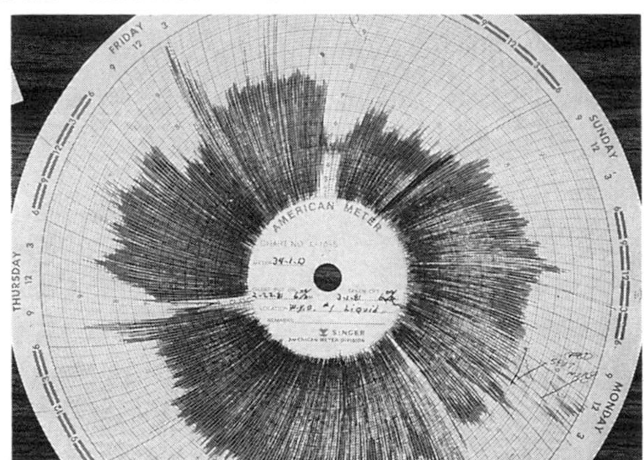

Figure 5-9 *Metering obtained with improper differential control.*

For example, an error of 0.5 inches at 100 inches represents a 0.23% error of flow, at 75 inches a 0.33% error of flow; but at 10 inches it creates a 2.5% flow error. The information shown in the following figure confirms these statements.The errors are shown for positive differential errors. A similar set of curves can be drawn for each case of low differential errors.

Good practice to achieve high accuracy dictates that the differential be kept as high as possible within strength limitations of the primary device, and the range of flow fluctuation be limited for the differential-measuring device.

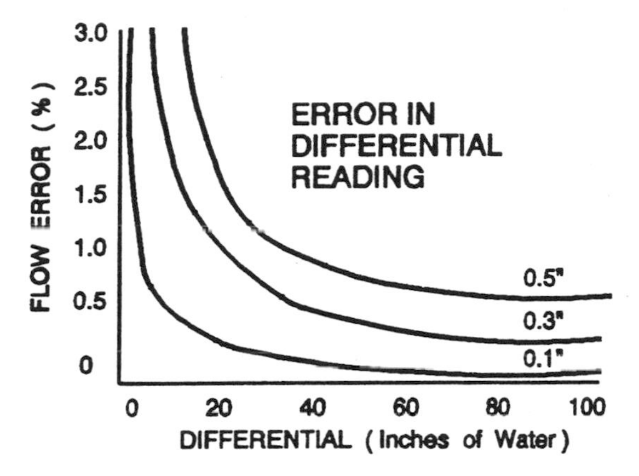

Figure 5-10 Percent error in flow vs. differential pressure with 0.1", 0.3", and 0.5" high zero.

Static Pressure An error in static-pressure with a head meter measuring gas is less significant at high pressure that at low pressure. The following figure shows this can become significant below 100 psig.

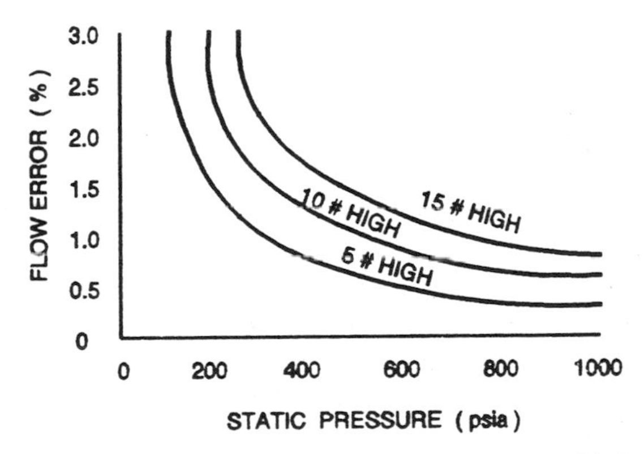

Figure 5-11 Percent error in flow vs. static pressure with 5, 10, and 15 psia high reading.

The differential error combined with error in static-pressure measurement at low pressure makes the head meter less accurate because of the expansion factor; it therefore has less range for low pressure, low differential in gas measurement. The only error for liquid is that due to the low differential.

Temperature Errors in temperature have a small effect on head meter flow accuracies for most gases, since the absolute ambient temperature range of measurement causes approximately 0.1% error in flow rate per degree Fahrenheit error — as shown in the Figure 5-12. For non-head meters, these temperature errors cause measurement errors twice as large.

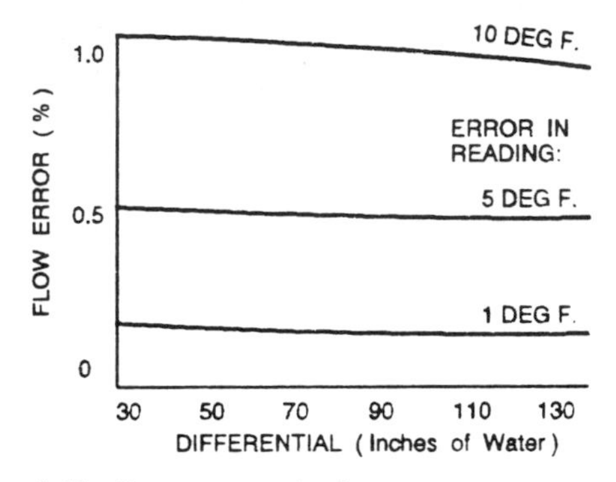

Figure 5-12 *Percent error in flow vs. temperature with 1°, 5°, and 10° F high reading.*

For liquids, the effects of temperature are much less. except for liquids lighter than water by 40% or more.

Specific Gravity Specific gravities of natural gas make about 0.1% error in flow with a head meter for each 0.001 error in reading; this can introduce fairly good-sized errors on gases with changing compositions unless this measurement is integrated into the volume calculation rather than averaged over a time period. This factor also enters into the "accuracy" statement of an orifice meter as a secondary factor in determining the compressibility factor. Specific gravity only affects the compressibility determination for non-head meters and does not enter as a direct correction.

For liquids, corrections for the effect of temperature and pressure are related to the measure of specific gravity or composition and must be used in the calculations.

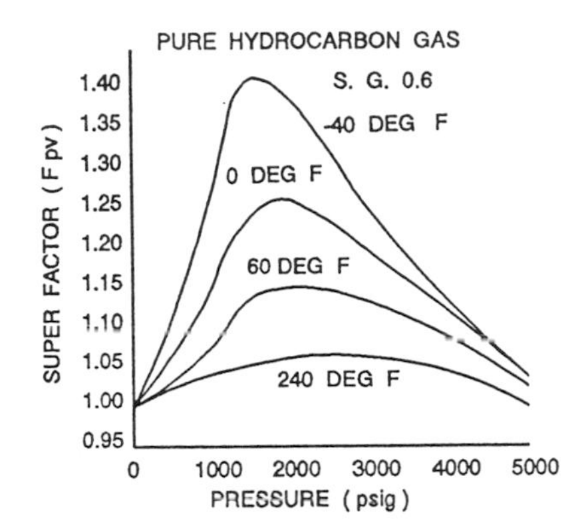

Figure 5-13 *Natural gas supercompressibility.*

Gas Compressibility The compressibility factor of natural gas (which corrects for the ratio of actual volume to ideal volume) is roughly a 0.5% correction in volume per 100 psi of pressure at usual measuring conditions on an orifice meter. Hence, an error of several percent in factor is only a small error in volume. However, if the gas is reduced near its critical point, correction factors as much as 225% are required, and small errors in measured variables are reflected as large errors of volume. These values are doubled for non-head meters.

Likewise, gases with large concentrations of non-hydrocarbon gases in their mixtures are difficult to calculate accurately since little data are available on these mixtures. Some of the theoretical values obtained by the pseudo-critical method (based on the mixture composition) have shown errors of several percent when compared with empirically determined test data on the same gas. This problem becomes more pronounced as the percentage of methane is reduced. If the value of the product handled is sufficient, then actual compressibility tests are recommended for confirmation of the theoretical data to the tolerances required.

Liquid Compressibility The compressibility factor for a liquid is not usually considered . However, at a flowing temperature within 75% (approx.) of the absolute critical temperature of petroleum, it must be considered.

If a specific weight device (commercially referred to as densitometer) is used, calculation of flow rate is simplified and the number of error sources reduced. Assuming an accurate device, the mathematical calculation of flow can be improved and the tolerance reduced. The usual four variables — temperature, pressure, specific gravity and compressibility — are reduced to one, density (if mass is being measured) or two, density and base specific gravity (if volume is the desired measurement unit.)

In each of the cases sighted for errors of the measured variables, there are two sources which cause errors: 1) measurement of the variables and 2) interpretation of the measurement for conversion to a mass or volume by use of the appropriate flow formula calculation.

Maintenance

It important that the meter be bought and installed with care, and equally important that a regular inspection routine be instigated to insure the "new quality" condition of the meter is maintained at all times.

If an inspector finds conditions existing that were not present originally and that cannot be corrected by cleaning, the meter should be removed from the line and rebuilt or replaced with a new meter. The importance of a regular meter inspection cannot be stressed enough, and the installation should be designed so that inspection is easily accomplished.

The primary element produces the signal to be received by the secondary element. As the secondary element cannot *improve* upon the signal produced by the primary element, all necessary care should be exercized in selecting construction materials and in maintaining both the primary and secondary elements.

Another source of error is the effect of time on the meter and the meter tube. No known pipeline is completely clean. The best that can be expected is a minimum of rust, solids, oil vapors, condensed liquid, lubricating greases, erosion and corrosion products, and the like. Any of these deposited on the meter and tube in the "right" places can cause errors of several percent. What this means to an operator is that the meter and the tube should be periodically inspected, cleaned and rechecked.

Figure 5-14 *Meters and meter tubes must be properly maintained for accurate flow measurement.*

Where sufficient money is involved, readouts are inspected once a month and meter tubes once a year.

Where there is less value being exchanged, tests may be made once every three months on the meters/readouts and every other year on meter tubes. However, the inspection time cycle should reflect the time period for the meter to get dirty and require cleaning. Periodic inspection will determine this time period.

Where sufficient solids (rust or sand) are present, slow erosion of the meter may require periodic replacement. Any corrosion pitting or buildup on the meter or meter run may require a meter or tube replacement. If there is sufficient foreign material, it may be necessary to install a filter or strainer in addition to periodic cleaning. However, cleaning usually takes care of this problem.

Recording and Calculating Data

Recording and calculating data is the final consideration for obtaining accurate flow measurement. All secondary devices must be calibrated against some standard. Likewise, when metering devices are exposed to widely varying ambient conditions, calibrations should be made covering the ranges encountered; if the effects are large enough, consideration should be given to controlling the environment in which secondary devices operate by adding housing with cooling and heating. Development of new smart transducers has given the user another option to take care of this problem — but at a higher price than for standard transducers. A balance between the accuracy required and the cost of obtaining it will determine the extent to which you can justify purchasing, testing and housing expenditures.

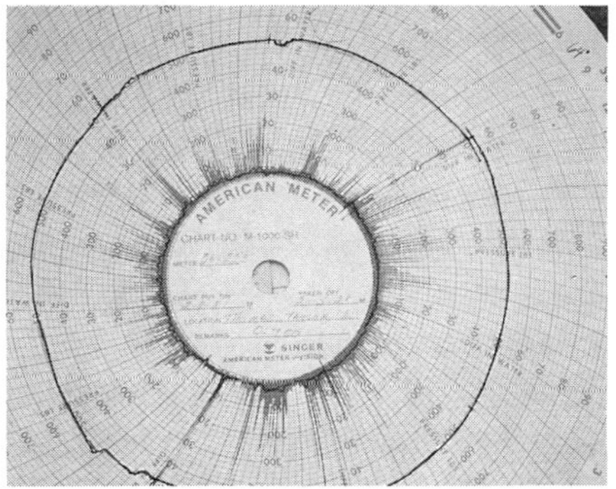

Figure 5-15 *Data recorded affects accuracy. The accuracy limit of this measurement would not be predictable.*

Indicated data must be either recorded and transferred to a central calculation office for conversion to flow rates or calculated directly by computer equipment installed at either location. Each step of recording or transducing and interpreting adds potential errors to the flow measurement, so a simpler system with proper maintenance has been found to yield the best results.

One of the most serious problems with the use of new types of recording and calculation equipment is the failure of manufacturers and users to recognize the need for proper training of personnel to get accurate measurement from such new equipment. Anyone who buys or sells equipment which is strange or different without an extensive training program can be assured that, at worst, the equipment will be found to be "no good" or at best there will be a period of time before familiarity is gained with the equipment and its capabilities realized.

Figure 5-16 *All new equipment must be understood if it is to be used with maximum effectiveness.*

Summary

By following guidelines in this book and procedures provided in the Standards and manufacturer's recommendations, primary elements can be provided that will offer the best possible accuracy in any specific measurement installation — provided all pertinent factors are remembered and considered.

Any overview of accurate flow measurement should contain a discussion about what kind of results can be obtained if all precautions are taken. Without full qualification of the source of the data and complete definition, accuracy numbers are meaningless. Proof of accuracy usually comes down to a study of system balances of measured flow inputs versus flow outputs. Experience tells us that the only way good balances are obtained is by following all of the best practices of design, application, installation, maintenance and interpretation.

There is no such thing as an absolutely accurate flow measurement. It is always done to some limit of accuracy. The purpose of any flow measurement should be to measure as accurately as possible within pertinent economic limits. For a flow purist, measurement should be as accurate as possible; but for the commercial flow measurement designer, the cost relative to the value of the measurement must be of concern. Since flow measurement has a tolerance, the designers job is to minimize this tolerance within the budget limits unless the value of more accuracy at additional costs can be demonstrated. Accuracy must be reaffirmable during flow for full confidence in a flowmeter's operation over time.

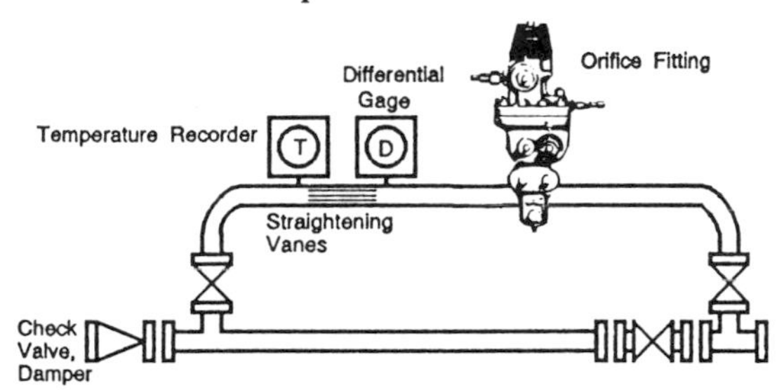

Figure 5-17 *A typical field meter station where costs are kept at a minimum.*

If a meter station is handling a product worth $1,000,000/day, a ± 0.2% inaccuracy is ± $2,000/day; this justifies investing considerable money to improve flow measurement. On the other hand, if the station is handling product worth $1,000/day, the same error of ± 0.2% represents only ±$2.00 per day — and this won't buy much improvement. There is also a law of diminishing return that applies to flow; beyond some point, additional expenditures will not buy better measurement.

References

1. American Chemical Society

2. ANSI/API 2530 (AGA-3) Orifice Metering of Natural Gas and Other Related Hydrocarbon Fluids. Arlington, VA, 1985.

3. American Petroleum Institute API Manual of Petroleum Measurement Standards. Publications and Distribution, 1220 L. Street N.W., Washington, D.C. 20005

4. ASTM Ordering Address: 1916 Race Street, Philadelphia PA 19103
Phone: (215) 299-5585

5. National Institute of Standards & Technology. 325 Broadway, Boulder CO 80303 — or Gaithersburg MD 20899.

6
FLUIDS

FLUIDS — LIQUIDS AND GASES

If the fluid is water at ambient conditions, then its influence on a flow measurement can be easily calculated from known and accepted data. However, if it is a fluid mixture near its critical temperature and critical pressure, then acceptable data may not be available, and the volume change with minor changes in temperature and pressure may make fluid definition the most important consideration in obtaining accurate flow measurement. *This is one of the most overlooked considerations in selecting and using a meter.*

A meter's commercially advertised accuracy normally allows for **no error** in determining fluid corrections, and users are misled into believing that simply buying an accurate meter will take care of all problems. If the fluid is not prepared for flow measurement, then no meter will provide "accurate" measurement. It is of value, therefore, to review the important fluid characteristics in order to know how to design an optimum metering system.

Good Flow Measurement Fluids

Good fluids:

1. are not near the flash point (for liquids) or condensing points (for gases)
2. are clean fluids with a composition whose PVT (pressure/volume/temperature) relationships are well documented with industry-acceptable data
3. are not exceptionally hot or cold since temperature may limit the ability to use certain meters
4. have minimal corrosive, erosive or depositing characteristics.

When these characteristics are considered and related problems answered, concerns are thus minimized over the influence of fluids on measurement accuracy, and metering is simplified.

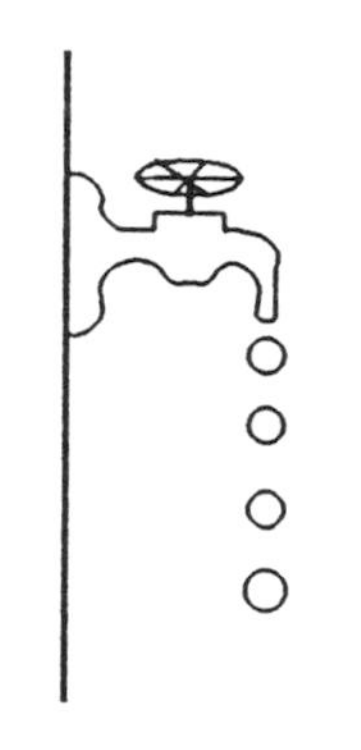

Figure 6-1 The water that comes out of your tap at home is a "good" fluid to measure.

Bad Flow Measurement Fluids

Many fluids classified as "bad to measure" become the main consideration in the choice of meter and the potential for measurement accuracy. Bad fluids include:

1. Two or more phases in the flowing stream
2. Dirty mixtures
3. Flows near fluid critical points
4. Flows with temperatures over 120°F or under 32°F
5. Highly corrosive or erosive fluids
6. Highly disturbed flows
7. Pulsating flows
8. Flows that undergo chemical or mechanical changes
9. Highly viscous flows

Specific meters may react differently to the problems listed above, and hence there may be one that works better than others for the specific problem presented. If should be recognized that the fluid sometimes must be measured even if it is a bad fluid and the cost of making it a "good fluid" is prohibitive in a cost/value study. The preferred conditions are sometimes simply not available at the point of measurement.

On the other hand, these characteristics may result in measurements with tolerances that are no better than ±20 to ±30%. An example is the measurement of carbon dioxide for injecting into oil reservoirs. Removing oil from the formation efficiently requires the CO_2 to be in the "dense fluid" stage near the critical point. Under these conditions — temperature and pressure near the critical points of 88°F and 1087 psia — CO_2 density variation may be one percent per degree Fahrenheit.

In this bad-fluid situation, the flow measurement designer would prefer to change the temperature or pressure or both. But, the use of the fluid precludes such change, so wider limits must be put on this measurement.

Basic Requirements and Assumptions

Fluid flow must be single phase for basic meter accuracy to be meaningful. There are two problems with a two-phase fluid. One is the effect on the meter, and the other is obtaining a truly representative sample to determine the composition for calculating the reduction to base conditions.

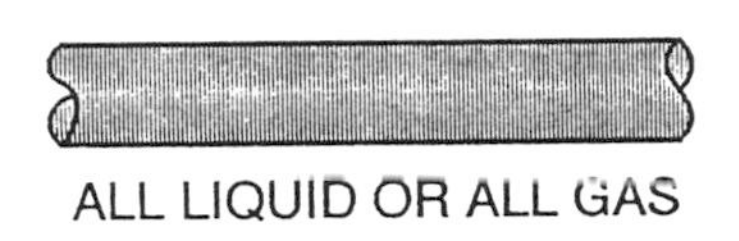

Figure 6-2 Single-phase flow is required.

For liquids in gas, the two-phase flow may affect mechanical configurations. Such flow patterns go through several regimes in a pipe line. The first phase tends to be droplets; if they are small enough, they form a homogenous mix so both the mechanical and sampling errors mentioned above are minimized and may not be significant.

The second regime comes when there are sufficient droplets to begin to accumulate in nonflowing areas including in the area of the meter where flow distortion may occur. Also, a sample can be inaccurate.

The third regime occurs with additional liquid. For example, there may be annular liquid flow with a core flow of gas. In this case, the mechanical configuration and the sampling are in trouble.

Adding even more liquid means there will then be two separate flowing streams (usually flowing at different velocities) forming layered flow.

The next regime occurs when slug flow occurs as liquids collect until the lines are filled to a point where they "burp" the liquids. There will then be no way to take care of mechanical or sampling problems since liquid slugs will be followed by gaseous pockets with no indication of which fluid is causing the meter to indicate flow. Separation must be done prior to attempting measurement in any of the cases where two-phase flow exists at the meter.

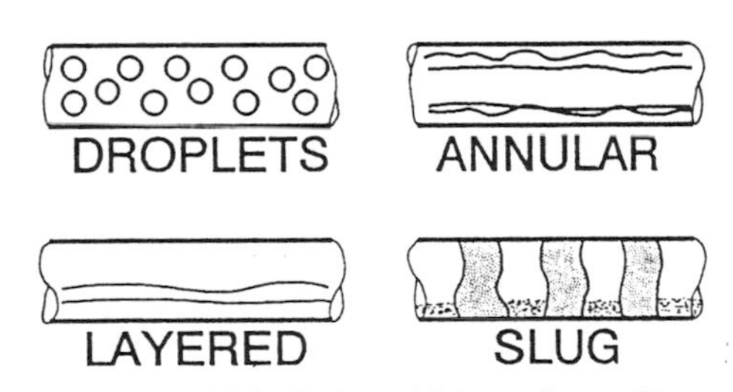

Figure 6-3 Disturbed, multiphase flow will produce inaccurate flow measurement.

In the reverse case— gas in liquids— similar problems exist, and deaerators must be used. With solids in liquids, filtering is required to allow correct measurement and minimize meter damage.

Some meters that react to mass can be successfully used for two-phase measurement provided they are not used to attempt to calculate volume without additional information on the fluid composition or density.

Some fluids are unstable and may present measurement problems by breaking down into other products during interrupted flow conditions or exposure to conditions along a line. Hydrates in natural gas measurement are an example. The hydrates, a mechanical combination of hydrocarbons and water that form an ice-like material, can block off main lines or gage lines so that readings of differential or static pressure are impossible. Likewise, crude oil can form an emulsion which does not lend itself to flow measurement. Some liquid plastics and such fluids as molten sulfur set up if they are not kept flowing or the meters heated. In these cases, fluids must be removed from the lines when shutdown occurs. The mechanical problems created may prevent measurement if these precautions are not taken.

Although problems caused by two phases have been outlined, some users are not aware that condensation of gases often may take place in a meter since this may be the lowest pressure and lowest temperature in the system. Whereas the fluid may be above saturation elsewhere in the system, conditions at the meter may be below the condensing point.

The same is true with flashing liquids; pressure within the meter may be lower than the pressure after going through the meter with recovery. If there is any question, excess back pressure should be planned rather than taking a chance on the fluid flashing.

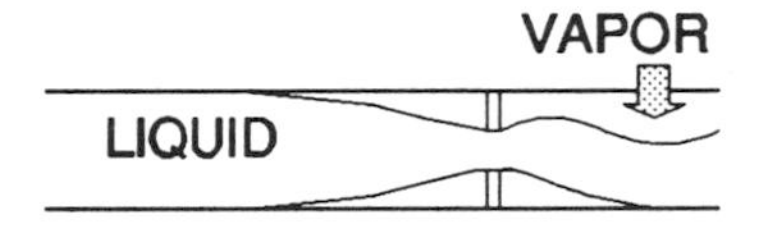

Figure 6-4 *Flashing fluids cannot be measured with accuracy.*

Fluids that are measured near their critical point have variations in PVT relationships such that the ability to measure pressures and temperatures within tolerances close enough to predict the effect on volume are beyond measurement equipment ability. Likewise, the accuracy of PVT correlation data deteriorates as the critical pressures and temperatures are approached. Because of these problems, flow measurement should not be attempted in such areas of operation if it can be avoided.

A number of meters have limitations for lower Reynolds numbers, and such limits should be checked prior to attempting to use a meter. The information on these limits is often quite general, and the user may be left in a quandary as to its meaning. Once again, a meter should be operated well away from such limits for best accuracy.

Summarizing: from a measurement standpoint, no fluid should be measured near a point of phase change, fluid characteristics change, near condensation, at too low a Reynolds number, or near critical pressure or temperature. Good flow measurement practice requires these conditions be recognized and proper precautions be taken.

Data Sources

Many data sources for fluid characteristics are available in the industry. The particular industry standards for specific applications should be the first place to look. Then there are general fluid specifications available from universities and from suppliers that handle certain products; manufacturers of fluid products have their own data. The "Flow Measurement Engineering Handbook"[11] has accumulated much useful data for general flow measurement.

Gas and liquids are usually considered as either pure products or commercial products in most references. Mixture laws may be used to estimate combined characteristics of uncommon fluids, but caution should be exercised if the mixtures contain widely varying molecular weights or involve extreme conditions of pressure and/or temperature. In

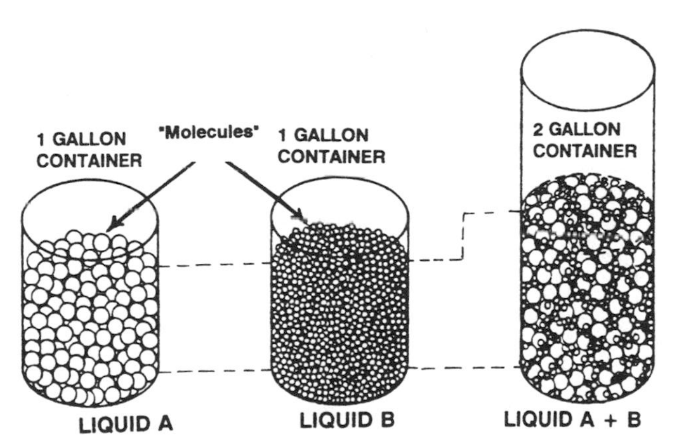

Figure 6-5 With marbles and "BB's" representing molecules, it is easy to see that mixing an equal volume of each does not result in twice the volume.

these cases, actual PVT tests should be run over the ranges of operation and a specific set of tables or equations set up for the particular mix, if such information can be correlated from the results obtained.

Listed below are some other sources. It is important to avoid the tendency to specify a mixed fluid by only its major component. Mixture characteristics must be considered, not just the PVT relationships of the pure product. Any data correlation must be examined to determine fluid data parameters involved before applying the data to a specific metering system.

Liquid propane is an example of a liquid that falls into this area of concern. Commercial propane contains 95% propane while percentages of the other constituents may vary; this variation will affect the correction factors used. Reagent grade propane, however, allows use of "pure" correction factors such as those available from the NIST (National Institute of Standards and Technology)[12].

Another way to approach the problem of variable fluid characteristics is to use densitometers or mass meters and measure mass. Then reduce pounds to an equivalent volume at base conditions based on an analysis and the density of the individual components at base conditions. (Note: this approach will be correct to determine a contract volume. But if there are significant differences in the flowing and base volumes, some means of deriving an approximate line-condition flow volume may be required for operations.) The contract volume is necessary for changing custody, but *actual* volume at line conditions is needed for operational control.

For any petroleum fluid, the API Chapter 11[1] of the "Manual of Petroleum Measurement Standards" is an excellent data source.

The ASTM[4] in conjunction with both the API[2] and the GPA[7] also put out common standards on petroleum-related fluids.

For international sources, the International Standards organizations have fluid data documents.

(See references 3, 5, 6, 8, 9, 10, 13, and 14 at the end of the chapter for other sources)

FLUID CHARACTERISTICS

Gases

The following discusses problems unique to some commonly measured gases.

Natural Gas is one of the most common gases measured since it is used both as a fuel and a feed stock in many industries. It represents the largest daily dollar volume of any gas routinely bought and sold. The natural gas industry has been the leader in developing gas measurement technology and its standards used in related areas of gas measurement.

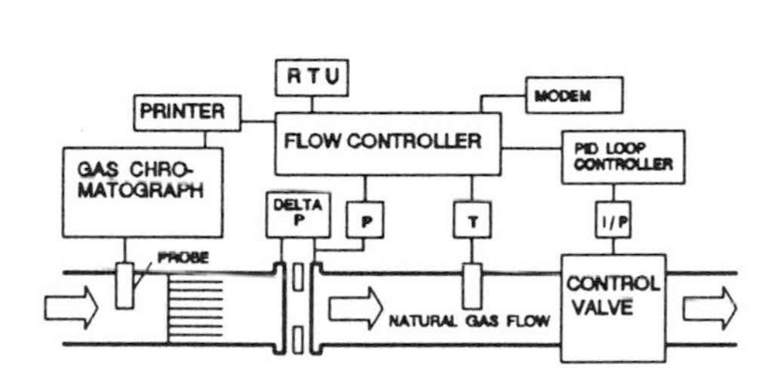

Figure 6-6 *Modern natural gas measuring systems may be quite complex, and all elements must be considered in determining metering accuracy.*

The ANSI/API "Manual of Petroleum Measurement Standards", Chapter 14, Natural Gas Fluids Measurement, Section 3, Concentric, Square Edged Orifice Meters Part 1, 2, 3, and 4 (also titled "American Gas Association Report Number 3," Parts 1,2,3,4, and "Gas Processors Association 8185-90" Parts 1,2,3, and 4) is the most common standard used for gas measurement of all kinds. Representing over 70 years of study of gas measurement with the orifice meter, the standard is continuing to be upgraded further by additional work.

Natural gas as a fluid varies from an easily measured fluid to a very difficult fluid to measure to close limits. Separated, dehydrated "pipeline quality" natural gas is normally easy to measure since it is a well defined fluid with very precise data available for relating the pressure, volume, and temperature from flowing conditions to base conditions. It is normally very clean with a minimum of solid "pipeline dust" and compressor or dehydration plant oils present. It is normally handled at temperatures and pressures that cause minimal meter design limitations or operating concerns.

Pipelines typically measure the gases within tenths of a percent in their pipeline balances between purchased gas, operating use gas, and gas sold.

Figure 6-7 *Flow measurement is at the heart of controlling many processes in modern refineries and chemical plants as well as for other industries.*

On the other hand, produced gas is often handled as a saturated fluid (separated to single phase but not dried), and the problems of flow measurement increase. The ability to balance a production field (with multiple wells) is thought to be within the industry norm if this balance is is within 3 to 5%. Other than some operating problems and poor maintenance that may affect measurement, the main cause of error is the fluid characteristics which cause both mechanical problems (liquid in the meter) and errors in fluid-density calculation (determining the proper specific gravity or relative density from a sample). Quite often, the volumes measured are not used directly for custody transfer, but for allocation in determining the percent of total volume contributed by each well's flow.

At the present time, measurement of two-phase fluid (gas and gas liquids and/or water) is not attempted because of the problems caused in a meter. There is a great deal of work being done on multiphase measurement which may, in time, result in acceptable two-phase measurement.

Getting a material balance in a system or processing plant can be one of the most frustrating flow measurement jobs. Careful attention must be paid to all of the concerns outlined here; even then, getting a close balance is difficult.

Mixtures of gases are more easily measured if the mixture has constant composition. This allows specific PVT tests to be run, or data may be available for common mixtures from prior work. The ability to use the mixture laws successfully has been previously discussed. If the mixture is changing rapidly, use of a densitometer or a mass meter may be required to determine an accurate quantity.

Ethane, a common chemical building block, may be measured as a pure product or as a mixture of an enriched stream from a processing plant. Data are available on ethane as a pure gas product. Except at conditions near the critical point, measuring ethane is a fairly straightforward metering problem.

However, its critical temperature of 90.1°F at a critical pressure of 667.8 psia may easily occur in normal pipeline and process plant operations. For the best measurement, some heating or compression may be required for conditions near critical to make them more favorable for measurement.

From a measurement standpoint, it is of value to remember that even though classified as a gas, ethane mixtures have characteristics that classify them as dense fluids up to approximately 1000 psia and at temperature from 90 to 120°F. In this area of operation (which quite often may exist in a pipeline or as part of a pipeline or processing measurement requirement), density changes significantly with small changes of temperatures and/or pressure. Because of this, flowmetering accuracy relates to these variables and proper selection of the equation of state for the ethane. If the temperature drops below 90°F, two-phase flow can be encountered at pressures below 670 psia. When either of these is likely, consideration must be given to adding heat or pressure to the system prior to attempting flow measurement.

Measurement of ethane as a pure liquid is not too common. Since the liquid has to be handled at a low temperature, it presents unique mechanical problems with meters. However, the data for making PVT corrections are available and accepted.

More common is the need to measure ethane liquid mixtures. As noted above, data on pure ethane are readily available, whereas data on mixtures with ethane-rich streams are limited, and accuracy will suffer accordingly. This is particularly true for variable mixtures. At times the better option is to use a densitometer with careful attention to proper sampling to minimize sample errors due to temperature and pressure variations from the flowing stream.

Propane can be handled as a liquid or a gas since its critical temperature is 206°F at a pressure of 616 psia. At normal ambient temperature, it can be a gas or liquid, depending on the pressure. However, to solve flow measurement problems, the phase relationship must be known so that a single phase of liquid or gas is considered. The meter can then be properly sized, and the two-phase or phase-changing regions can be avoided. As with any fluid, the closer to phase change the measurement is attempted, the more difficult the measurement becomes.

Ethylene is a popular hydrocarbon feed stock used in the chemical industry. It is difficult to measure in a number of industrial cases since its critical temperature is 48.5°F at 731 psia which means that it is handled in an area of density sensitivity (i.e., the measurement problem becomes one of the correct density measurement). Near this point in the vapor phase, the compressibility factor changes as rapidly as 1% per degree Fahrenheit and 0.5% per pound pressure. This problem is significant enough to, at times, require heating prior to flow measurement to obtain accuracy. Heating to 80°-90°F will minimize these density changes so they can be handled. This represents an example where the fluid characteristics are significant enough to the measurement accuracy that they are changed prior to attempting measurement.

Another characteristic of ethylene is that in its processing, a very fine carbon dust is produced that cannot be removed with a 5 micron filter. Depending on the frequency of buildup, the meter and meter piping must be cleaned so that flow characteristics are not changed. This also affects measurement transducers and their lead lines which must be cleaned.

Ethylene often contains small quantities of hydrogen. This can affect filled differential transducers; an internal pressure is built up over a period of time to the point that the unit will not be capable of meeting a calibration, and replacement is required. How often this happens varies with pressure and hydrogen content.

Ethylene requires special elastomer materials for meter internal parts that come in contact with the fluid ("wetted" parts). Meter materials should be checked, and meters should be ordered to accommodate necessary requirements.

Propylene is another popular feed stock for the chemical industry. It is somewhat easier to measure than ethylene. This is true because the critical temperature is 197°F at a critical pressure of 667 psia. Propylene is less reactive with most meter materials, but materials in meter seals should be checked carefully for reaction with propylene.

Carbon Dioxide is commonly measured in industry and used in the oil and gas industry for recovery of crude oil. It also has a troublesome critical temperature of 88°F at a critical pressure of 1071 psia that is significant in attempting flow measurement. The compressibility factors in these ranges

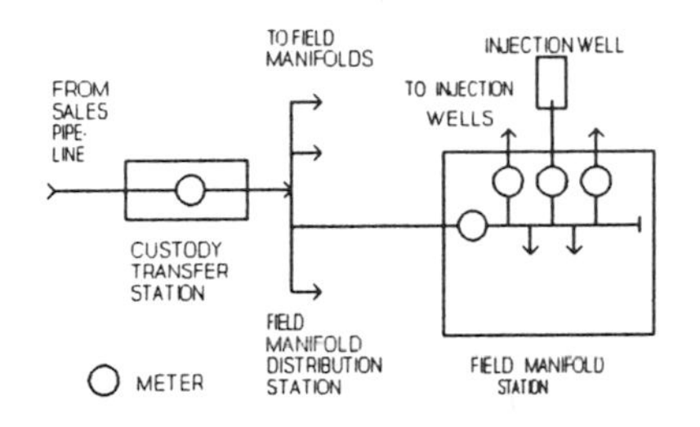

Figure 6-8 Typical CO_2 injection system for crude oil recovery.

may represent as much as a 200% correction and will be the controlling factor in achieving accurate flow measurement. Carbon dioxide is quite often handled as a mixture which further complicates the density determination. Data are available from the National Institute of Standards and Technology (NIST) in Boulder, Colorado for mixtures of CO_2 from 94% to 99.7% containing small amounts of methane (0- 2%), ethane (0-1%), propane (0-2%), and nitrogen (0-2%), as well as pure CO_2.

Once again, carbon dioxide measurement is not easy because of the large changes in density at normal operating conditions— even if well removed from the two-phase flow region.. The solution requires continual integration of density with the flow device because of the rapid changes. A computer is needed to do this calculation rapidly enough. Another measurement problem is CO_2 wetness. If water content is present to a sufficient level, a hydrate may be formed at temperatures well above freezing (32^{0}F). In addition, wet CO_2 is very corrosive, and a large amount of corrosion products will move with the gas; the result is often deposits which can cause problems with meters and other equipment such as densitometers. Likewise, CO_2 causes most standard seal materials such as rubber and Teflon to break down. Lubricants rapidly deteriorate in the presence of CO_2. All these factors should be taken into account before designing a carbon dioxide metering station.

Steam flow measurement is probably the poorest flow measurement made in industry. There are may reasons for this, but the fluid problems are the most important.

The measurement of steam as a fluid is fundamentally the same as the measurement of any flowing fluid. If the fluid dynamics and the thermodynamics are known, then the first steps towards accurate measurement have been taken. However, these two areas are not particularly well known or understood in many of the applications where steam measurement is a necessity.

As commonly used, the term "steam" is meaningless when considering measurement. In the industrial world, the definition narrows somewhat. But even here the term steam covers a variety of flowing conditions. The following sections cover the three possible flowing conditions where steam is being measured: wet, saturated, and superheated.

Wet (Quality) steam is usually the most difficult fluid to measure. A wet steam is a fluid that contains both condensed hot water and steam. In the two-phase portion of the phase diagram (see Figure 6-9), with the same temperature and pressure there is a different density. Therefore, a third parameter, quality, must be added to correct a measurement for the right density. Quality is defined as the percent of flow that is steam and water by mass fraction. For example, 95% quality means the 95% is steam and 5% is water by weight.

Quality can be determined by a calorimeter test (batch operation) which is good only until changes in the system take place (i.e., flow rate change or density change). In addition to these fluid definition problems, the effect of the two phases on the meter mechanism can create errors. Because of the foregoing, quality steam measurement is inaccurate and should be attempted only as a last resort, recognizing that it will have much wider accuracy tolerances.

Saturated steam has no water present and exists only at one pressure and corresponding temperature. At the same pressure and a higher temperature, the steam is *superheated*; at a lower temperature, condensation takes place and the fluid becomes quality steam. Saturated steam exists at a boiler. But when steam leaves a boiler (assuming no superheat is added), the flow creates a pressure drop and there is a possibility of a temperature drop depending on insulation of the lines. Therefore, steam traveling through a plant will normally be superheated (i.e., the pressure drops, but the temperature is relatively constant and the steam is almost never saturated away from the boiler).

Measurements required to determine steam density:

	T	P	Quality
Wet	x	x	x
Saturated	T or P	T or P	
Superheated	x	x	

Most designers state that they will be handling "saturated steam" and may not allow for temperature and pressure measurement at the meter. As the flow rate varies, the pressure will change at the meter, and sometimes the temperature will also change. To determine density, temperature and pressure must be known (measured). If true saturated steam exists at the meter, then measuring temperature or pressure will define the density.

Wet steam requires a measure of temperature, pressure and quality to define density. Saturated steam requires temperature or pressure, and superheated steam requires temperature and pressure. In each case, these measurements must be fed to a computer that calculates the density based on the equations in steam tables.

A problem unique to steam is the large difference between the temperatures of ambient air surrounding a meter and the flowing fluid. This makes the proper measure of steam temperature a major concern. Without suitable insulation and special precautions, major errors in density will result. In a flowing stream at low velocity, steam tends to stratify by temperature; mixing of the stream is necessary to get the temperature constant across the stream and allow accurate measurement of the temperature — and hence density measurement.

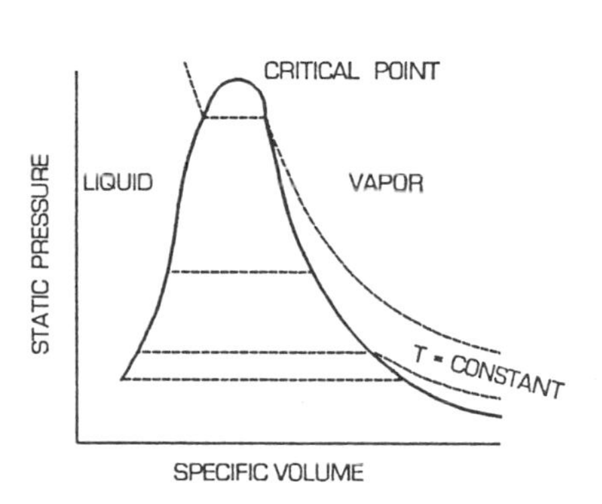

Figure 6-9 Steam phase diagram.

In summary, steam is the most difficult fluid to measure accurately. Even with the utmost care, a plant balance of steam and water flows in a power-generating or process plant is very difficult to do well.

LIQUIDS

Liquids are generally reputed to be easier to measure than gases. That might be true if all liquids were like water at ambient temperatures. They are not, however, and there are certain precautions for each of the fluids covered below that need to be considered for the best flow measurement.

Crude oil has had as much research done on it as any liquid in terms of product value as well as the worldwide measurement and exchange of the product. The generic term "crude oil" covers a multitude of fluid variables that can be categorized by the following terms: light, medium, heavy, dirty, sweet, sour, and waxey. These conditions all affect flow measurement.

Light crude is the most desirable from a measurement viewpoint. Its viscosity range and wax content are the lowest, and both of these effect metering. Some heavy crude cannot be measured without heating. The high wax content creates the likelihood of deposits in lines and meters. When this occurs, no meter can operate properly. Wax treatment or heating is required before measurement is attempted. As previously discussed, flowing temperatures must be known to define the magnitude of the measurement problems. A statement about light crude must be accompanied by the operating temperature range to allow proper metering system design.

Sweet and sour crude oils typically affect meter materials and involve foreign material present due to corrosion. Knowing this, corrosive-resistant meters can be selected, and/or corrosive products can be filtered out prior to the meter.

Most meters have a sensitivity to viscosity which limits the range some can handle. Adding to the problem is the effect of temperature on viscosity and the fact that the temperatures of crude oil measurement cover wide ranges and are normally not controlled, but instead are determined by the situation. For example, storage-tank oil may run 120^0 F, whereas tanker oil arrives at ocean temperature, and pipeline flows typically arrive at ground or ambient temperature.

Medium and heavy crudes have intermediate characteristics, but high viscosities and "crud" content can increase the metering problems.

Complete data available in the literature has been agreed upon by petroleum measurement groups worldwide. These data were accepted in early 1980 (in the U.S., August 1980) and are referenced by contract. The data make corrections from flowing conditions to base conditions and cover not only crude oil but all liquid hydrocarbons within the defined data base limits.

Dirty crude includes foreign material which will collect in a meter. It is experienced in some production areas where the term "grass" is generally applied to the phenomena. It should be removed by filtering prior to metering or it will stop the meter. Once pipeline quality crude is achieved, this problem seldom occurs.

Refined products, as the name indicates, are processed so that foreign materials have been eliminated. There are normally limited variations in constituents since refined products must meet product specifications established by the industry. A large quantity of industry data is available (as mentioned in the crude oil section). For new products, the individual buyers and sellers will develop accepted data based on PVT tests covering the ranges of operation. Then, when the product becomes widely traded, the industry will correlate the various data, run additional tests as necessary, and publish industry correlations. Hence, most of these fluids have well defined characteristics which can be used with confidence to get good measurement.

Ethylene and propylene liquids are well defined if they are pure products. Unless they are handled near their critical temperatures, they introduce no specific problems. Concerns of measurement near the critical points of these two fluids, or for their mixture problems when they are not pure, have been pointed out previously in the comments about each.

Gasoline is easy to measure since it is stable in the temperature ranges at which is it normally handled and has no tendency to flash to a gas. Correction factors are well established and accepted, and they are available in many meter readout systems with no special programming required. There are no viscosity or foreign material problems to cause special concern about a meters operation.

Heavier hydrocarbons (i.e., C_{10}^+ and heavier) present potential problems of viscosity that must be addressed since they are normally handed in smaller quantities and small meters have more problems with viscosity effects. These fluids also may be unstable at normal handling temperatures and require controlled temperature and pressure conditions.

Natural gas liquid mixtures produced with natural gas contain variable amounts of the light hydrocarbons (ethane through pentane). They are significant fluids in the oil and gas industry and have received a great deal of attention for flow measurement. As indicated in their definition, natural gas liquids are undefined mixtures that quite often change composition with time, pressure, and temperature. They are recovered from separators whose efficiencies are related to operating temperatures and pressures which change between night and day. The range of variations in flowing temperatures and pressures can be wide, and this further complicates the measurement. With the wide range of molecular weights of some of the components (ethane 30 and pentane 72), mixtures of the products have variable shrinkages with changing compositions.

The problems above have given rise to a number of individual operating company tables and tables available from the Gas Processors Association[7] that correlate a restricted data set. With this problem of large changes in parameters, many operators have opted to use a densitometer combined with a volume flow meter to measure mass flow, or a true mass meter which measures mass rate. A true mass meter does not require the densitometer. If an analysis is done, the mass flow can be converted to volume flow by knowing the cubic feet per pound at base conditions of the individual components. The quantity of mass or volume required will define equipment required. With wide variations in fluid characteristics, the procedure of using mass and analysis is the most accurate way of measuring these flows, particularly at extreme temperature or pressure.

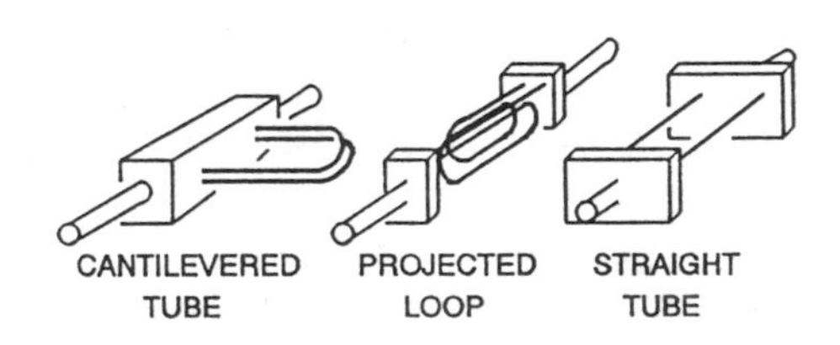

Figure 6-10 *Types of Coriolis true mass flow meters.*

When natural gas liquids are processed, they can be broken down into their pure components which are more fully defined liquids. However, it is very important that quality requirements for the products be known since they affect flow measurement. If the fluid is reagent grade ethane, propane, or butane, then industry accepted pure-product correlations may be used for correcting from flowing to base conditions.

On the other hand, if commercial grade products are specified, then mixtures of hydrocarbons are allowed, and the correlations of the pure products must be adjusted to reflect corrections based on the mixtures' specific gravities. Here again, industry accepted tables may be used within the limits of their data bases.

When a customer states he wants to measure "propane," the meter system designer does not know exactly what is required. Reagent grade has over 99.5% propane, but commercial propane requires only 95% propane. The amount of ethane and butane may vary and cause correction factors to vary. So-called propane-rich streams may have even <u>lower</u> propane percentages. Ethane-propane mixtures present a measurement problem with the fluid PVT relationships, since normal handling conditions are near critical conditions.

There are numerous tables from operating companies based on their own data bases. These are available from the Gas Processors Association[7]. There is also a study to standardize the temperature correction values in liquid. The specific gravity range of 0.35 to 0.70, with a temperature range of roughly 50°F to 150°F, generally covers the EP mix. The purpose of this study is to compare all standard procedures and, if possible, pull the industry together to agree on a single relationship. (Availability was imminent when this book was being publsihed.)

Two-phase flows fall into two general categories with most meters: measurable and unmeasurable. Since current techniques do not always provide the ability to prevent two-phase flow, studies have been made of ways of handling the problem in limited ranges. Within these specified limits, the methods have been correlated based on the density of the two individual streams so as to address the problem of up to 5% by volume of gas in liquids and up to 2% by weight of liquids in gas. These are very low limits and should not be applied to higher-content mixed flow. In each of these cases, the accuracy tolerance of such measurement is at least double that expected of single-phase measurement with a given meter. These procedures have been applied to steam and condensed water systems, natural gas with natural gas liquids, crude oil and gas flows.

Some true mass meters can measure two-phase flows within design limits. Flows outside of these limits are unmeasurable and should be separated and measured as individual liquid and gas flows.

References

1. American Petroleum Institute API Manual of Petroleum Measurement Standards. Chapter 1, Physical Properties Data.

2. API Manual of Petroleum Measurement Standards. Chapter 14 — Natural Gas Measurement, Section 3 "Concentric, Square-edged Orifice Meters," Parts 1, 2, 3, and 4.

3. ASME Steam Tables in "Thermodynamic and Transport Properties of Steam, 2nd edition. New York, N.Y., 1967.

4. American Society of Testing & Materials, 1916 Race Street, Philadelphia, Pennsylvania 18103

5. Benedict-Webb-Rubin Equation of State in Journal of Chemistry & Physics. 8,334, 1940.

6. Fischer & Porter Catalog 10-A-94. Warmister PA., 1953. (Contains liquid specific gravities and viscosities.)

7. Gas Processors Association, 1812 First Place, Tulsa OK 74103 Phone: (918) 585-5112.

8. Gas Processors Suppliers Engineering Data Book (Handbook). Volume II, 10th Edition, 1987, Tulsa OK

9. Hankinson and Thompson Corresponding States of Liquid Densities. 1979

10. Matheson Gas Products Matheson Unabridged Gas Data Book, "A Compilation of Physical and Thermodynamic Properties of Gas. 1974

11. Miller, Richard W. Flow Measurement Engineering Handbook. New York: McGraw-Hill Book Company, 1983.

12. National Institute of Standards & Technology, 325 Broadway, Boulder CO 80303.

13. National Institute of Standards & Technology NBS Technical Note 1045, "Density Algorithm for Ethylene. Boulder CO 80303: NIST, 1981.

14. Starling, Kenneth E Fluid Thermodynamic Properties for Light Petroleum Systems. Houston TX: Gulf Publishing Co., 1973.

7
FLOW

Flow of fluid can aid or detract from the ability to make a measurement. Certain basic assumptions made previously in this book are amplified below.

Required Characteristics

The required characteristics of the flow include: continuous, nonfluctuating, nonpulsating and the pipe running full.

Continuous means that the flow should not continually be off and on. Each meter has a certain amount of inertia to start, plus overshoot after a flow stops. Furthermore, during these periods, the inaccuracies of the low rate measurements are well below quoted meter accuracies.

On the other hand, start-up and stop requirements for flow measurement are common for batch type operations. No meter measures correctly from zero flow to normal flow. However, when totalized flow is the desired goal, short startup and shutdown times can be insignificant to a total flow, provided they are a small percentage of the total flow time.

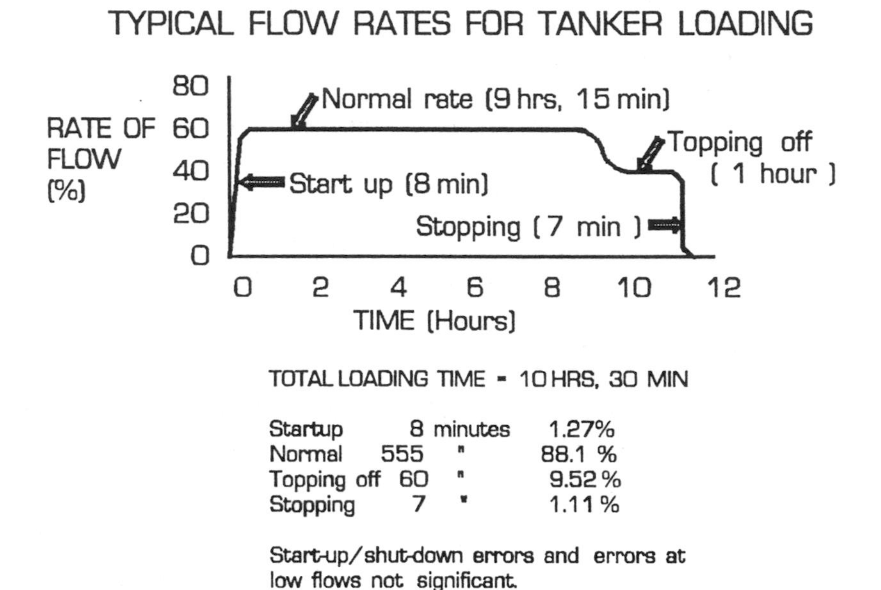

Figure 7-1 *Short periods of inaccurate flow measurement can be unimportant when very large volumes are involved.*

For example, loading a supertanker may take 12 hours, and a start of 8 minutes may be needed to get up to loading rate. Of this 8 minutes, it may be less than two minutes before a meter gets into its accurate range. This represents less than a few tenths of the total loading time and less than a hundredth of the total fluid loaded — usually compatible with inaccuracies of the measuring system.

On the other hand, if a large portion of the metering time and/or much of the total flow is at low rates, then a multiple meter system should be designed. For large swings the system might have one normal meter, one meter to measure at the low rate and a third to handle an occasional high peak. The key design parameter should be the percent of the total to be measured that is represented by excursions of the flows.

When on-off flows operating a control system are important at each rate of flow, then multiple-flow-range systems *must* be used.

Fluctuating flow really represents a flow rate change at a rate below the frequency response of the flow system. If the rate exceeds the frequency responses of the system, the metering may appear to continue but the accuracy can be severely compromised. Operators may have dampened readout systems to the point that the records look good, but the accuracy has been destroyed. So much dampening can be applied that any meaningful signal to the readout system is obliterated and an apparently absolutely constant rate results. Such a rate has virtually no relationship to fluid actually flowing. The proper method of correcting for a fluctuation of flow is to dampen the variation in the stream flowing in the pipeline through the meter to below the meter's frequency response. If this is not possible, then a meter with faster response may be required to obtain measurement.

Pulsation versus fluctuation is a matter of the frequency of flow changes. Pulsations may also come from pressure changes not related to flow but which can cause errors. In either case, the complexity of the problem strongly suggests that these variations in flow or pressure be removed before flow measurement is attempted. The problem is more prevalent in gas than liquid measurement.

Most metering systems commercially available do not have a frequency response high enough to respond to flow pulsations more rapid than a few hertz at best. The effects from a pump, compressor, control valve, or flow-created pulsations may exceed meter frequency limits. Here again, improper dampening of the readout system may make an operator happy with the flow record, but can introduce major flow readout errors of over 100%. Since recognizing pulsation significant enough to cause error in flow is very difficult, several instruments have been developed to help predict whether or not pulsation is causing flow-measurement problems. None of these are capable of being used as a factor correction device; they are used only to discover pulsation and/or to show that the pulsation has been eliminated sufficiently to allow valid flow measurement.

Proper piping design to minimize acoustical "tuning" in different meter systems can be a useful approach to minimizing problems in both the primary meter piping as well as in the secondary instrumentation. Experience has shown that a majority of the flow measurement problems with head meters caused by pulsation occur in the secondary instruments. Short, large-diameter piping to the differential meter is recommended since such piping tunes only to very high frequencies (over 100 hertz) which are normally above the frequency that causes flow error. When pulsation is present, all statements of meter system accuracy are suspect. The pulsations should be eliminated before flow measurement is attempted.

Full-conduit flow is important in liquid systems. The flowing pipe must run full, or measurements made will be in error. This can be a problem if piping design does not keep the meter below the rest of the piping. If the meter is at the high point, then vapor can collect and create a void in the meter so any velocity or volume displacement measured will be in error.

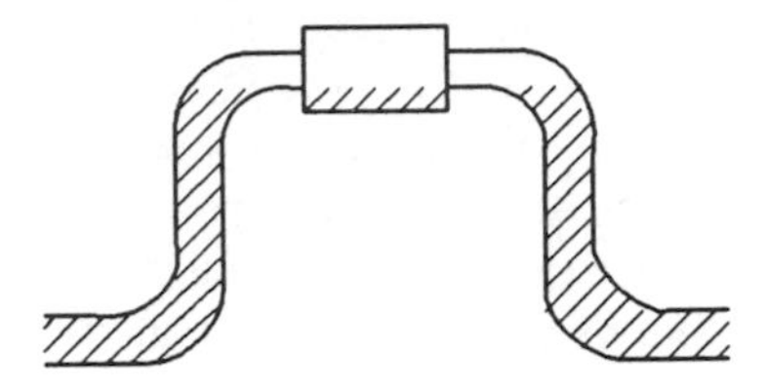

Figure 7-2 *To keep the pipe running full with liquid flow, do not place the meter higher than piping. For gas flow, keep the meter higher than piping.*

Measurement Units

In the measurement of flow, output is desired either as some unit of volume or some unit of mass. Volume without a definition of base conditions does not represent a definite quantity. With a base set of conditions, the terms of volume and mass are equivalent in representing a defined quantity and relate to each other through density at base conditions. The most common method to measure either value as a volume or as a mass is to use a densitometer that represents the density at line conditions. Mass meters may be used to measure mass directly. If volume is desired from a mass measurement, it can be calculated from a density of the fluid at base conditions (usually calculated from a component analysis).

The proper base value to be used is defined by the plant contracts, government requirement, or agreement. The base conditions control the choice of metering system, particularly the readout.

Installation Requirements

Each meter has certain requirements for installation into a pipeline. The requirements vary from one type of meter to another. Refer to the specific meter section for details of these requirements. The requirements have been determined by standards organizations, users, manufacturers and research laboratories. In each case, the performance characteristics of the meter depend on these requirements being met. Any lessening of the specifications means that performance is compromised, and specific calibrations in place should be run.

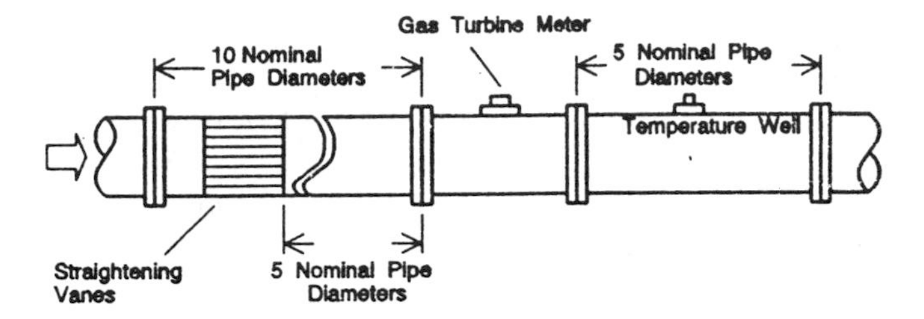

***Figure 7-3** Recommended installation of an in-line gas turbine meter (minimum lengths) — as per AGA-7*[1].

Most meters have been developed from some basic principle embodied in a prototype. These prototypes may go through several iterations. But if they prove out in a flow test, they become a marketable product. Except in special cases, most meters are designed for general usage and are tested accordingly. During these tests, effects on meter performance from flow into and out of the meter are determined and installation requirements set. With sufficient experience in a given industry, plus additional tests, a standard may be prepared to guide users on the meter's installation requirements.

As time passes and additional applications are checked, these requirements may be adjusted to reflect new knowledge. Whatever its source, data should be checked to assure a user that the application has been evaluated and the design data are valid for the specific job requirements.

Flow Pattern

The heart of all this is the flow pattern into the meter which, in general, will be correct if the Reynolds number limit and the inlet piping agree with the original evaluation. The installation will then be proper. If a designer deviates from the tested design, most meter manufacturers make no claims on their meters' performance and simply state that minimum piping requirements must be met.

This means two things are important: 1) familiarity with installation requirements is a must, and 2) the minimum requirements must be met to make sure no errors are introduced due to installation. Some meters have a defined flow pattern requirement (i.e., the amount of distortion of the flow pattern and/or swirl allowed), but it is rarely checked when a meter arrives in the field. Recent work has indicated that additional flow-pattern preparation is necessary to reduce the inaccuracies in meters directly sensitive to velocities. In the future this may require changes in installation requirements to maintain proper flow patterns.

Much of the significant research into installation requirements has been conducted by the AGA-API[1,2] orifice meter committees responsible for writing the standard on orifices. This work basically applies to gas. Since Reynolds number correlation indicates valid results with gas, the principles should work even better on liquids, and these designs should thus be conservative. The best interpretation of this requirement is to use the worst case (most disturbed profile) and design to it with the highest Reynolds numbers (i.e., largest beta ratio) so the tube will be of a universal length that is independent of the actual piping layout upstream or meter tube length. This will require longer meter tubes than the shortest that may be used under the standard, but all lengths listed in the standard are the minimum required with longer tubes recommended if space allows. Expanding this to other meter applications will yield lengths longer than used on many applications, but will provide a conservative design approach.

The International Standards Organization (ISO)[3] also has a standard written on the same orifice meter subject. It is in conflict to a degree with the AGA-API[1,2] standard in that it requires roughly twice the length that is required in AGA[1] for meter runs without straightening vanes, and much longer lengths for meter tubes with straightening vanes. The standard states that if these lengths are used, no widening of accuracy tolerance will be required. The optional shorter lengths included for tubes without vanes are similar to the AGA lengths. But if they are used, an additional 0.5% must be added to the tolerance of the volumes metered.

These discrepancies, as well as questions by U.S. users about some of the AGA lengths, have caused additional studies to be made that seem to fall somewhere between the requirements of the two standards when straightening vanes are used. The discrepancies are being worked on and may be redefined in the next few years. In the meantime, as the desire for reduction of tolerances on flow measurement continues, there seems to be a tendency to require longer lengths for utmost accuracy.

Flow pattern distortion comes from improper upstream pipe fittings, foreign material in the meter tube, or improperly made or aligned meter tubes. Any of these will cause an asymmetrical flow pattern with the possibility of swirl. Asymmetrical profiles can cause errors in the 0-5% range. Swirl can cause errors in the 10-50% range with differential meters and generally less with other type meters.

Obviously, swirl is the major concern. If there is any question of swirl being present, then straightening vanes should be used. Experience has

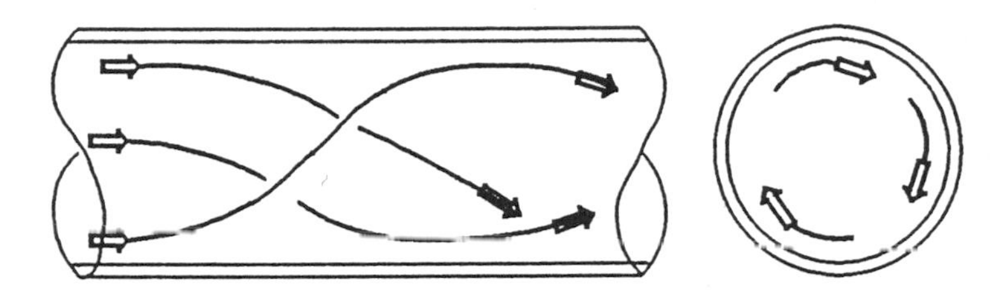

Figure 7-4 Swirl must be eliminated for accurate flow measurement.

indicated that swirl can propagate even in extra-long upstream pipe. The better design is to minimize changes in direction and planes of flow upstream of the meter. If swirl is not generated, it won't be propagated.

In addition to generating swirl, piping upstream of the meter can create or minimize other problems. High velocities across blocked-in tees to lines can create vortices in both the tee and the line. This can set up fluctuations or pulsations that make flow metering difficult. On gas lines that may drop out liquids, pockets of liquids can also create variable flow rates as the fluids wash back and forth. Drips with drains should be put in low spots to allow drainage of this liquid.

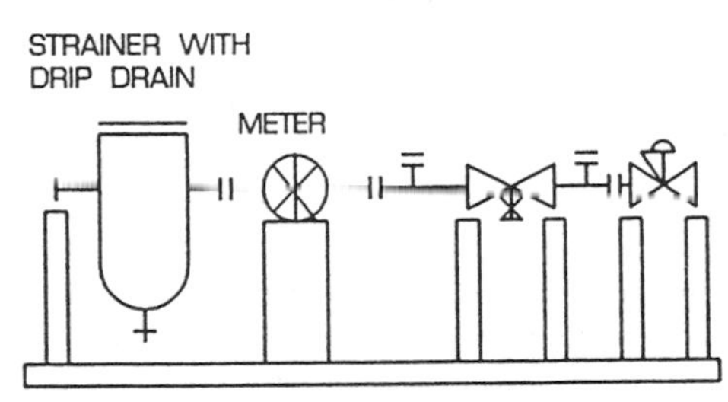

Figure 7-5 Drips help keep gas lines free of liquids, and air eliminators keep liquid lines free of gas.

If pipelines are very dirty, it may be necessary to filter fluids out of gas and solids out of liquids. On liquid lines, where gas or vapor are present, deaerators should be installed or flow fluctuations may result. Sizing of upstream headers should be controlled so that velocities are slowed in the lines leading to meter runs. A rule of thumb is the area of the header should be 1½ to 2 times the area of all of the meter runs off of it. Another rule-of-thumb is to have velocity in the headers one-half of that in the meter tubes. This will provide good flow distribution to the meters and minimize flow profile problems.

References

1. American Gas Association, 1515 Wilson Boulevard, Arlington VA 22209

2. API Manual of Petroleum Measurement Standards. Chapter 14 - Natural Gas Measurement Section 3 "Concentric, Square Edged Orifice Meters" Parts 1,2,3, and 4.

3. International Standards Organization, Case Postale 36, CH 1211 Geneve 20, Switzerland — or in US: ANSI (for ISO Standards), Customer Service, 11 West 42nd Street, New York, NY 10036 Phone: (212) 642-4900

8
MEASUREMENT AND METERS

After considering — and taking care of — all effects of fluids and flow to and through a meter, meter selection and application can be completed. Several general considerations should be reviewed as well as specific characteristics of individual meters.

METER CHARACTERISTICS

Comparing Meters

Characteristics to be considered when evaluating a meter include: accuracy, comparative cost, use acceptance and use, repeatability, maintenances costs, operating cost, few or no moving parts, ruggedness, service life, rangeability, available in style to meet fluid property problems, available in pressure and temperature ratings required, ease of installation and removal, power required, pressure loss caused by meter (running and stopped), and how well calibration can be proved.

No single meter will have all of the characteristics desired, but candidates can be evaluated by going through such a list for each meter under consideration and then deciding which of the factors are of prime importance for the particular flow measurement problem. A procedure something like the following may be helpful:

1. List, for each candidate meter, the characteristics of importance.
2. Define *how* important each is by assigning a weighting factor (such as from 10 for very important to 0 for no importance).
3. Assign a similar rating number to show how well the meter performance meets the specific characteristic.
4. Multiply the weight factor by the performance factor, and add all the totals.

Comparing totals for various meters will provide a rank ordering. (This same concept can be applied to the total metering system problem by including considerations for flow, fluid, installation, maintenance, etc.)

MAXIMUM FLOW CAPACITY OF GAS MEASUREMENT METERS (CU. FT. / HR.)

(At 14.7 psia and 90 degrees F.)

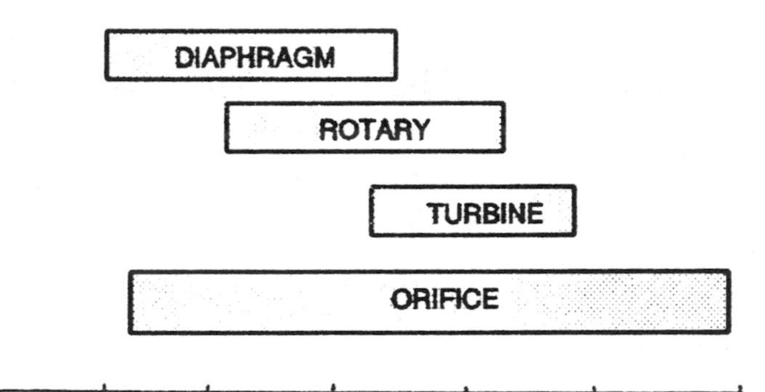

10 100 1000 10,000 100,000 1,000,000 10,000,000

CFH OF NATURAL GAS AT BASE CONDITIONS

Figure 8-1 *Comparison of flow ranges for various gas flowmeters.*

There are no industry required standards or testing facilities where flowmeter tolerances can be validated and a ranking of meters published. Even if such a facility were available, it would be impossible to make a test expressing all possible uses of a meter. A simple case in point: a test on gas may not relate to a liquid metering problem or vice versa, even with consideration of the Reynolds number effect. Therefore, specific information may not be available, but knowledge *is* available in the industry to give *guidance* on the probabilities of success for a given meter in a given application.

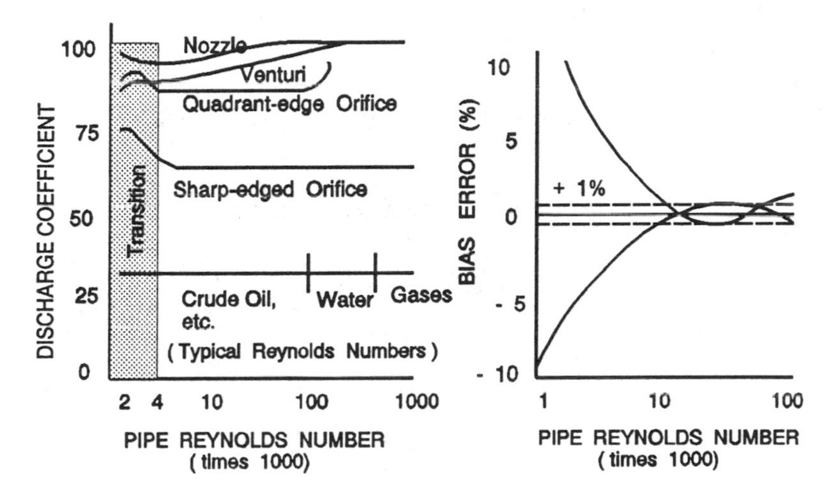

Figure 8-2 *Shown here are typical effects on Discharge Coefficient for various types of meters and Bias Error versus Reynolds number in measuring typical fluids.*

Specific meters prove useful somewhere to flow metering — or they quietly disappear from the marketplace. The flow measurement fraternity is a relatively conservative group. New meters based on a new principle have a long road to acceptance under the best of circumstances. The history of every meter in use today has followed this road. A meter must provide acceptable performance/cost ratios, considering initial cost plus the required investment in proper installation, operation, and maintenance, or it won't complete the trip down the acceptance road.

The chart on the following pages overviews the major flow meters and salient characteristics. This is by no means a definitive chart since it lists only general categories and performances. For a specific set of flow conditions, the comments may or may not apply. Once again, reference should be made to relevant industry data, manufacturers' information, and the experience of others to evaluate flow metering for a given application.

Text continued on page 111

FLOW METER CONSIDERATIONS

METER TYPE	COSTS				PIPE SIZE (Inches)	
	INSTAL-LATION	OPER-ATION	MAINTE-NANCE	WITH READOUT	LIQUID	GAS
Orifice						
Sq-Edge Concentric	Med	Med	Lw-Med	Lw-Med	0.5-100	1.5-30(1)
Honed Meter Tube	Med	Med	Med	Med	0.25-1.5	0.25-1.5
Eccentric	Med	Med	Lw-Med	Lw-Med	2-30	2-30
Segmental	Med	Med	Lw-Med	Lw-Med	2-30	2-30
Quadrant Edge	Med	High	Lw-Med	Lw-Med	2-8	NA
Conic	Med	High	Lw-Med	Lw-Med	2-30	NA
Flow Nozzle	High	Med	Lw-Med	Med	2-30	2-30
Ventturi	High	Low	Lw-Med	High	2-30	2-60
Turbine	Med	Med	Md-Hi	Med	2-30	2-30
Positive Displacement	Md-Hi	Lw-Med	Md-Hi	Md-Hi	2-30	2-12(2)
Magnetic	Md-Hi	Low	Md-Hi	Md-Hi	0.1-100	NA
Vortex Shedding	Med	Med	Med	Med	0.5-8	0.5-8
Ultrasonic						
Doppler	Lw-Med	Low	Lw-Med	Lw-Med	>0.5	>0.5
Transit-Single Path	Lw-Med	Low	Lw-Med	Lw-Med	>0.5	>0.5
Transit-Multipath	Lw-Med	Low	Lw-Med	High	>4	4-42
Coriolis	Med	Md-Hi	Med	High	0.125-6	0.125-6
Pitot						
Single Port	Low	Low	Low	Low	>1.0	>1.0
Multiport	Low	Low	Low	Low	>1.0	>1.0
Rotameter	Low	Low	Low	Low	0.125-24	0.125-24
Target	Med	Med	Med	Med	0.5-4	0.5-4
Elbow	Low	Low	Low	Low	2-30	2-42

(Low, Medium, High)
(NA = NOT APPLICABLE)

(1) Standards limit
(2) Varies with type of PD meter

NOTE: THIS CHART IS A *GENERAL* GUIDE.
In special situations, variations can be found for almost every entry in this chart.
It is a broad overview of major meters; some special meters have been left out since they are either similar to meters shown or have very limited use.

APPLICATION/SERVICE									
Liquids					Gas		Steam		
CLEAN	DIRTY	VIS-COUS	CORRO-SIVE	SLURRY	CLEAN DRY	DIRTY WET	SATUR-ATED	SUPER-HEATED	QUALITY
Rec	Ltd	NR	Ltd(1)	NR	Rec	Ltd	Rec	Rec	Ltd
Rec	NR	NR	Ltd(1)	NR	Rec	NR	Rec	Rec	Ltd
Rec	Rec	NR	Ltd(1)	NR	Rec	Ltd	Rec	Rec	Ltd
Rec	Rec	NR	Ltd(1)	NR	NR	Ltd	Rec	Rec	Ltd
Rec	Ltd	Rec	Ltd(1)	NR	NR	NR	NR	NR	NR
Rec	Ltd	Rec	Ltd(1)	NR	NR	NR	NR	NR	NR
Rec	Rec	NR	Ltd(1)	NR	Rec	Ltd	Rec	Ltd	Ltd
Rec	Ltd	NR	Ltd(1)	Ltd	Rec	Ltd	Rec	Ltd	Ltd
Rec	Ltd	Ltd	Ltd(1)	NR	Rec	NR	NR	NR	NR
Rec	NR	Ltd	Ltd(1)	NR	NR	NR	NR	NR	NR
Rec	Ltd	NR	Rec	NR	NR	NR	NR	NR	NR
Rec	Ltd	Ltd	Ltd(1)	NR	Rec	NR	Rec	Rec	Ltd
Rec	Ltd	Ltd	Rec(1)	Rec	NR	NR	NR	NR	NR
Rec	Ltd	Ltd	Rec	NR	Rec	NR	NR	NR	NR
Rec	Ltd	Ltd	Rec	NR	NR	NR	NR	NR	NR
Rec	Ltd	Rec	Ltd(1)	Ltd	Ltd	Ltd	Ltd	Ltd	NR
Rec	Ltd	NR	Ltd(1)	NR	Rec	NR	NR	NR	NR
Rec	Ltd	Ltd	Ltd(1)	NR	Rec	NR	NR	NR	NR
Rec	Ltd	NR	Ltd(1)	NR	Rec	NR	NR	NR	NR
NR	Rec	Rec	Ltd(1)	NR	Rec	NR	NR	NR	NR
Rec	NR	NR	NR	NR	Rec	NR	Ltd	Ltd	NR

(Recommended, Not Recommended, Limited)
(NA = Not Applicable)

(1) Special materials required.
(2) Varies with type PD meter.

| • • • • • • • • • • CHARACTERISTICS • • • • • • • • • |

METER TYPE	ACCURACY Proved	ACCURACY Unprov'd	RANGE-ABLT'Y	REYNOLDS NUMBER (Min)	PIPING REQR'D (Diam ")	OUTPUT	PRES. LIMIT (psig)	TEMP. LIMIT (°F)
Orifice								
Sq-Edge Concentric	±0.25	±1.00	3/1(Note 1)		>4000	10-80	Sq Rt	(Materials Note 2)
Honed Meter Tube	±0.50	±1.00	3/1(Note 1)		>1000	10	Sq Rt	(Materials Note 2)
Eccentric	±0.50	±2.00	3/1(Note 1)		>4000	10-40	Sq Rt	(Materials Note 2)
Segmental	±0.50	±2.00	3/1(Note 1)		>4000	10	Sq Rt	(Materials Note 2)
Quadrant Edge	±0.50	±2.00	3/1(Note 1)		>4000	10	Sq Rt	(Materials Note 2)
Conic	±0.50	±2.00	3/1(Note 1)		>4000	10	Sq Rt	(Materials Note 2)
Flow Nozzle	±0.50	±1.50	3/1(Note 1)		>10000	10-40	Sq Rt	(Materials Note 2)
Ventturi	±0.50	±1.50	3/1(Note 1)		>7500	5-30	Sq Rt	(Materials Note 2)
Turbine	±0.25	±3.00	(See Note 3 below)		5-20	Linear	3000	-450/+500
Positive Displacement	±0.50	±1.00	(See Note 4 below)		None	Linear	(See Note 4 below)	
Magnetic	±0.50	±1.00	(Note 5)	No limit	5-10	Linear	750	360
Vortex Shedding	±0.50	±2.00	(Note 6)	>10000	5-40	Linear	3600	750
Ultrasonic								
Doppler	±2.00	±5.00	(Note 7)	No limit	5-20	Linear	1000	-300/+200
Transit-Single Path	±1.00	±3.00	(Note 7)	No limit	5-20	Linear	(Note 2)	-300/+500
Transit-Multipath	±0.50	±1.00	60/1	No limit	6	Linear	(Note 2)	-300/+500
Coriolis	±1.00	±5.00	(Note 7)	No limit	None	Linear	3000	-300/+400
Pitot								
Single Port	±1.00	±5.00	$3/1^{(1)}$	>4000	20-30	Sq Rt	(Materials Note 2)	
Multiport	±1.00	±5.00	$3/1^{(1)}$	>4000	20-30	Sq Rt	(Materials Note 2)	
Rotameter	±2.00	±5.00	10/1	>100	None	Linear	(Note 3)	(Note 8)
Target	±1.00	±3.00	$3/1^{(1)}$	>100	10-20	Sq Rt	(Materials Note 2)	
Elbow	±0.50	±4.00	$3/1^{(1)}$	>10000	10-40	Sq Rt	(Materials Note 2)	

Note 1: Single differential pressure
Note 2: Special materials required
Note 3: Rangeability 10/1 liquid, 100/1 gas
Minimum Reynolds Number 3000 liquid, 10000 gas
Note 4: Rangeability 20/1 liquid, 250/1 gas
Minimum Reynolds Number >100 to >1000
depending on type meter

Note 5: 10/1 to 100/1
Note 6: 10/1 to 30/1
Note 7: 5/1 to 25/1
Note 8: See Note 3, but
varies with type meter

FLOW METER CONSIDERATIONS

METER TYPE	● ● ● ● ● MAJOR FACTORS IN USE ● ● ● ● ● Limitations
Orifice	
Sq-Edge Concentric	3/1 range, high pressure loss, subject to upstream conditions
Honed Meter Tube	Must calibrate small sizes, must keep clean
Eccentric	3/1 range, high pressure loss, subject to upstream conditions
Segmental	3/1 range, high pressure loss, subject to upstream conditions
Quadrant Edge	3/1 range, high pressure loss, subject to upstream conditions
Conic	3/1 range, high pressure loss, subject to upstream conditions
Flow Nozzle	Hard to inspect, sizing must be right, expensive
Ventturi	Hard to inspect, sizing must be right, expensive
Turbine	High maintenance cost. See also Note 1 below
Positive Displacement	Limited temp/pres, flow stops when meter stops, bulky, heavy, needs clean flow.
Magnetic	Does not operate on hydrocarbons
Vortex Shedding	No swirl (input piping), low resolution for large meters
Ultrasonic	
Doppler	Clean fluids required, upstream piping required to condition flow
Transit-Single Path	Clean fluids required, upstream piping required to condition flow
Tranist-Multipath	Clean fluids required
Coriolis	Sensitive to vibrations, not available above 6"
Pitot	
Single Port	Needs full flow profile, limited sampling, low delta-P
Multiport	Must mount vertical, remote readout needs special equipment
Rotameter	Only small sizes available
Target	Low delta-P for normal flow
Elbow	Custom manufacturing/calibration, low delta-P for normal flow

Note 1: On gas, sensitive to density and friction effects;
on liquid, sensitive to viscosity effects.

METER TYPE	Advantages
Orifice	
Sq-Edge Concentric Honed Meter Tube Eccentric Segmental Quadrant Edge Conic	Well established, users confident, same readout on all sizes, wide capacity with plate changes, no calibration, easy to install, low cost, no limits on materials, no power required, largest database of user information of all meters.
Flow Nozzle	High capacity, same readout for all sizes, no sharp edges, self-cleaning
Ventturi	High capacity, same readout for all sizes, no sharp edges, self-cleaning
Turbine	High frequency readout, linear calibration, wide temp/pres operating range
Positive Displacement	Best for viscous fluids, does not need pipe straighteners, simple readout
Magnetic	No head loss, bidriectional, no effect from density or viscosity
Vortex Shedding	Wide rangeability, no moving parts, measures many fluids, easy to install
Ultrasonic	
Doppler	Handles slurries, clamp on existing lines, no pressure drop
Transit-Single Path	No pressure drop, bidirectional
Tranist-Multipath	No pressure drop, bidirectional, limited upstream piping, reads irregular profile
Coriolis	Measures mass flow, handles some two-phase flows
Pitot	
Single Port	Low pressure drop, same readout for all sizes
Multiport	Low pressure drop, same readout for all sizes
Rotameter	No special upstream piping
Target	Handles dirty flow, uses common readout
Elbow	Same readout for all sizes

Flow Meter Specification Terms

Continued from page 105

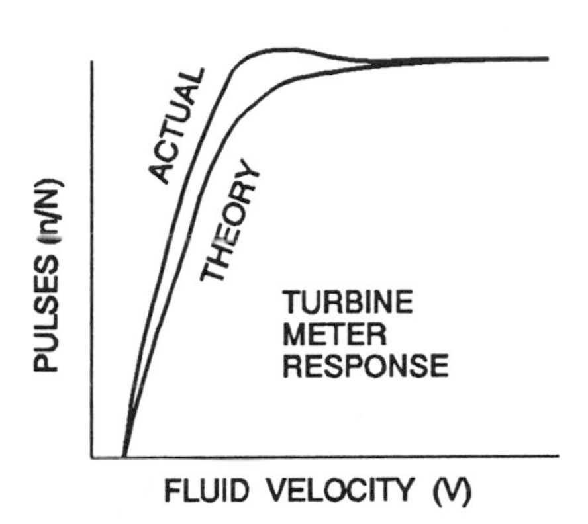

Figure 8-3 Many factors besides accuracy specifications are important for a flow meter performance curve. For example, meter response speed may depart from theory.

Meter accuracy is a much-abused term. Manufacturers often default to it for "selling shorthand," but those specifying and buying meters to measure flow should be aware of several caveats.

In the first place, no meter is *absolutely* accurate. There are no *absolute* standards of gas or liquid against which to compare a meter reading to see how closely that reading compares with what is *actually* flowing through the meter. Furthermore, any statement of accuracy must include not only the best possible estimate of how inaccurate the measurement is but also over what flow range the estimate applies. The more diligent manufacturers usually supply such detailed information. But that is only part of the problem in determining meter accuracy.

As previously noted, it is equally vital to know and take into account the type of fluid being measured, the conditions under which the meter will be used (including fluid condition), how it will be installed, and what level of maintenance will be provided. Otherwise accuracy statements are meaningless in terms of values actually obtainable.

Rangeability expresses the flow range over which a meter operates while meeting a stated accuracy tolerance. It is often stated as "turndown" — maximum flow divided by minimum flow over the range. For example: a meter with maximum flow (100%) of 100 gallons per minute and minimum flow (within a stated tolerance such as ±.5%) of 10 gpm has a 10-to-1 rangeability or turndown of 10. It will be accurate ±0.5% from 10 to 100 gpm.

The meter may provide a tighter tolerance over a more limited range — say 10-to-1 within ±0.5% of actual flow and 3-to-1 within ±0.25%. This means the user can select the tighter tolerance of ±0.25% for a range of 33% to 100% flow (33 to 100 gpm).

Linearity defines how close to a specific accuracy the meter registers over a stated flow range. Its proof curve will approximate a straight line. It may be significantly inaccurate but quite linear.

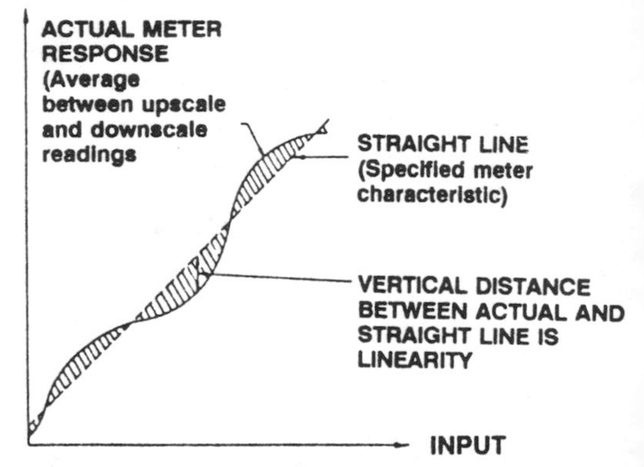

Figure 8-4 *Linearity describes how well a meter "tracks" the theoretical accuracy curve.*

It is important to point out that this characteristic was much more important prior to the widespread use of computers and electronic signal-conditioning equipment. With a computer correction device, it is possible to characterize a non-linear meter output curve *provided the meter output is repeatable.* Such curve characterization allows a closer fit of the readout system — even for a linear meter — to minimize calibration errors. The same procedure is used on "smart" transducers to minimize any non-linearity the transducer may exhibit as a result of temperature and/or pressure effects.

Repeatability means just what it says — how nearly the same reading a meter will provide for a given flow condition. As with linearity, it may be more important to always get the same reading for specific flow rates than that those readings be extremely accurate. Flow control is an example of this need.

Hysterisis is closely related to repeatability. It describes what happens to meter output as a given flow rate is approached from larger and smaller flow rates. For example, suppose a flow rate of 80 gpm is increased to 100 gpm, and a meter then registers 99 gpm. Now the flow rate increases to 120 gpm and returns again to 100 gpm; the meter registers 101 gpm. Its hysterisis is ±1 gpm, and the dead band is 2 gpm at a 100 gpm flow rate.

Consideration of these meter characteristics shows clearly that simply relying on a manufacturer's statement of accuracy is indeed an incomplete and inadequate way to evaluate and compare meters. And, as stated, proper determination and application of accuracy, rangeability, linearity, repeatability, and hysterisis data is basic but still only part of the job in achieving the best flow measurement; operation and maintenance must also be considered.

Meter Factor is a correction that mathematically modifies a meter's indication to a corrected "true" reading based on knowledge of the flow and flowing conditions. Corrected readings may be manually calculated periodically or the meter factor automatically applied continuously. It is normally determined from a throughput test, covering the range of flows to be measured, based on a master meter or a prover.

TYPES OF METERS

There are many ways that types of meters can be listed; they are listed here by their individual names under general categories, since there are significant differences between the meters in the same category (i.e., differential head meters of the orifice, flow nozzle, venturi and pitot tube types). They are listed in random order, and there is no significance to the order.

Head Meters

For several centuries, the basic concepts of head meters have been known: flow rate is equal to velocity times the device area, flow varies with the square root of the head or pressure drop across it. Likewise, the equations of continuity were well known.

The first two systems designed on these basic concepts were the Pitot (1732) and the Venturi (1797). The flow nozzle was used in the late 1800's, and the orifice appeared in commercial use in the early 1900's. In each of these cases, the original investigators set their own requirements of design configuration, calculation, installation and operation of their units. Continual research on and use of the various head devices since this early work have resulted in standards for the construction, installation, operation and maintenance of them. Research continues today, reflecting the continuing interest in the head meters as useful flow metering devices.

The head device consists of a primary element that causes a pressure drop related to velocity with specifications on the geometry of the differential producer, the length and condition of the adjacent piping and the pressure tap locations. It is of primary importance to remember that to take advantage of a head meter's predictable calibration from its mechanical

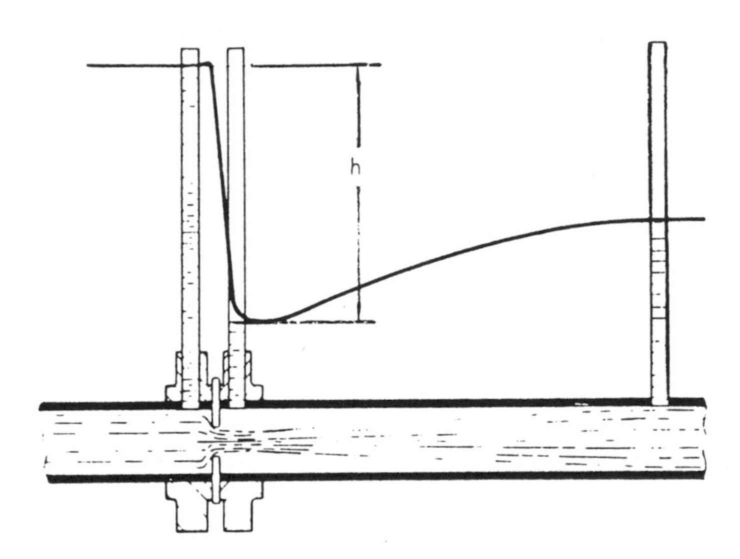

Figure 8-5 Typical orifice flow pattern and pressure differential across the orifice.

construction, these specifications must be known and followed. Any deviation from these requirements necessitates a calibration to determine the relationship of the pressure drop to the flow rate.

The secondary device consists of a differential pressure measuring unit, connecting piping, and other measuring units required to define the flowing variables of the fluid such as pressure, temperature, and composition.

The advantages of a head meter can generally be listed as follows:

1. simple
2. inexpensive
3. calibration inferred from mechanical construction
4. available in most sizes
5. easy maintenance
6. rugged
7. widely used and accepted
8. does not require power
9. can be built with special materials.

Specific head meters may not have all of the above advantages, but these are the general considerations for choosing a head meter. For example, the Venturi and the pitot meters have relatively low permanent pressure losses.

Others, such as the orifice, are adaptable to capacity changes by changing plates provided the capacity change is not rapid with time. The Venturi and the nozzle will handle dirty fluids.

The head devices have few limitations in terms of application and are used for most types of flow measurement. They have been particularly applied in the measurement of water, natural gas, hydrocarbons of all kinds, chemicals, etc. It is easier to list where they are not used, such as: viscous fluids, slurries, pulsating fluids and non-Newtonian fluids. Even these limits can be handled with special care under some circumstances.

In summary, the head devices cover a large category of flow meter in service today. Their wide acceptance and use is based on successful applications and service over many years.

Orifice Meter

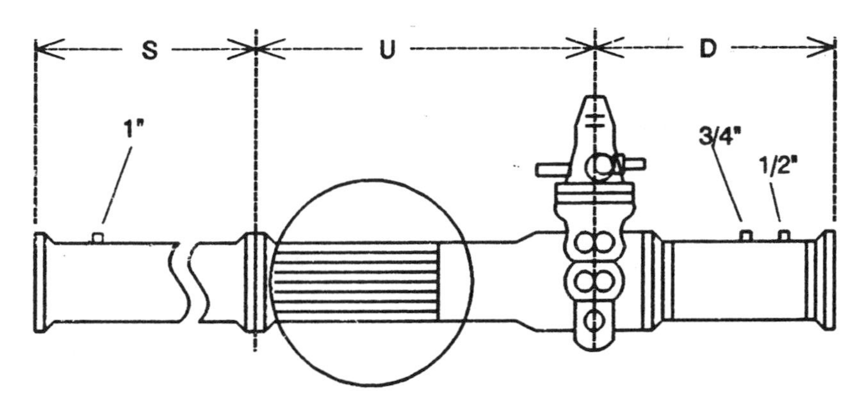

Figure 8-6 *Typical orifice meter built to AGA-3 specifications.*

Construction of the orifice meter is extremely important since its mechanical construction defines its calibration under the various standards written on the subject. One of the most complete standards written on the orifice is the API, Chapter 14, Natural Gas Fluids Measurement[5] — more commonly called AGA-Report No. 3[1].

AGA-3 is the "bible" of the orifice meter. It represents the compilation of many different tests covering over 80 years with the latest agreed-upon knowledge of: construction and installation, method of computing flow, orifice meter tables for natural gas, nomenclature and physical constants, and appendices. It is subject to periodic review and is updated as new knowledge is gained. It represents the most widely used standard for high pressure natural gas measurement and is successful in commercially measuring the majority of the natural gas exchanged in the world.

AGA-3[1,5] is also used to define an orifice installation for liquid flows. Its usefulness has been found to be directly related to the knowledge of its contents. Introduction to the standard is covered here with the hope that this will form a basis for further study to properly complete the job of measuring with an orifice meter. The standard is based on a very large number of tests and experiences of the practitioners of flow measurement; to use the report, conditions found to be significant must be reproduced.

The basic premise of the standard is to make it unnecessary to calibrate each individual meter and be able to predict and control its measurement accuracy by controlling the way it is made and installed. This has been found possible to a tolerance acceptable for commercial measurement, especially for head devices.

The most important of the primary element head devices is the orifice. It comes in many configurations for special applications, but by far the most widely used is the sharp-edged, concentric orifice plate using flange taps (located one inch from the upstream face of the plate and one inch from the downstream face of the plate.)

Reynolds numbers and beta ratios are the keys to accuracy limits of these plates. With a beta of roughly 0.10 to 0.75, the percent of uncertainty is 0.5 percent or less. However, at Reynolds numbers below 1,000,000, uncertainty is increased as shown here:

Reynolds number	Uncertainty (%)
100,000	0.53
50,000	0.55
10,000	0.70
5,000	0.70

Uncertainty becomes so large — several percent — for Reynolds numbers below 4000 that practical measurement is no longer possible. The lowest Reynolds numbers typically occur on small runs with small plates.

An orifice used on liquid is limited by the lower pipe Reynolds numbers listed above which are often more important because of the higher viscosities of liquids. The reason is the nonlinearity of the flow coefficient with slight changes in the Reynolds number; at these low ranges the large changes in flow coefficient cause significant errors in flow-rate measurements unless coefficient, volume and Reynolds number calculations are iterated continually. For accurate measurement, rangeability of an orifice with a 100" recording device is usually limited to 3-to-1 for a single installation. However, multiple differential readout devices or multiple tubes are commonly used to extend the orifice rangeability.

The major orifice use is for general-purpose, non-viscous flows where low cost is important. Compared to some other meters, it has a relatively high permanent pressure drop for a given flow rate, which may limit its use if pumping costs are a major consideration. Maintenance of the primary device consists of a periodic inspection and cleaning of the plate and the meter run since inaccuracies can occur if initial equipment condition is not maintained. The primary advantages of the orifice use are its wide acceptance, simplicity, the large number of trained operating personnel available to maintain it, and the large amount of industry research data available on it.

The orifice standard: AGA[1,5] (Latest edition dated 1992) AGA-3 usefulness depends directly on how well the information in the 1992 standard is applied. Subjects covered by the standard include:

Part 1 - General Equations and Uncertainty Guidelines
Part 2 - Specification and Installation Requirements
Part 3 - Natural Gas Applications
Part 4 - Background, Developments and Implementation Procedure and Subroutine Documentation for Empirical Flange-Tapped Discharge Coefficient Equation.

Part 1 of AGA-3 contains the basic flow equations for all fluids to be used to calculate flow rate through a concentric, square edged, flange-tapped orifice meter. It provides an explanation of terms used and methods for determining fluid properties. A definition of the uncertainty and guidelines for calculating possible errors in flow of fluids are given.

Part 2 of the standard gives specification and installation requirements for the manufacture and use of orifice plates, meter runs (upstream and downstream) orifice plate holders, pressure taps, thermometer wells and flow conditioners. These specifications and requirements must be followed in detail to use the calculations of Part 1. Variations from the requirements on initial installation as well as during use of the station will create the potential of error in flow measurement. In-place testing is required to determine what calibration factors should be applied since the error estimates of the standard do not apply to non-standard metering situations.

Part 3 is specific to the use of the standard for natural gas measurement as practiced in North America.

Part 4 discusses the background, development, and limitations of the coefficient of discharge data. This material is of academic interest but is not required to apply the data. However, the second section of Part 4 is important because example calculations detail the procedures to implement the data. The purpose of this section is to minimize accounting differences obtained by various computers and/or programmers in the applying data.

The latest version of the new document is different in some details from previous standards in the equations used and the requirements for manufacturing and installing an orifice. A grandfather clause allows stations built under previous standards to continue to be used. The new standard contains the latest, most defensible data, and it reduces the uncertainty of measurement over a wider range of application of the orifice for flow measurement.

Other Standards that cover the use of orifices include ISO-5167[8], ASME-MFC-3M[7], and ANSI/API 2530[5] (identical to AGA-3) documents. Since most of them trace back to the same basic data bases, there are similarities in the documents. However, there are differences that must be understood since an orifice built to one standard may not meet the other standards.

The measurement fraternity is attempting to resolve these differences. In the meantime, if a specific standard must be met, then it must be specified in ordering and operating equipment.

Orifice Meter Description

An orifice meter consists of an orifice plate, a holding device, upstream and downstream meter piping, and pressure taps. By far the most critical part of the meter is the orifice plate, particularly the widely used square-edged concentric plate whose construction requirements are well documented in standards such as AGA-3[1,5] and ISO 5167-1[8]. These standards define the plate's edge, flatness, thickness — with bevel details, if required — and bore limitations.

The most common holding system is a pair of flanges; however, for more precise measurement various fittings are used. These usually also simplify plate installation/removal for changing flow ranges and for easy inspection. In every case, the orifice must be installed concentric with the pipe bore within limits stated by the standard.

An orifice plate installed without specified upstream and downstream lengths of pipe controlled to close tolerances and/or without properly made pressure taps (usually flanges) is not a legitimate flowmeter; it must have specific tests run to determine its calibration. Since this is not economical, orifice systems are built to meet the standard(s); this allows calculations to be made with specified tolerances. Control over orifice metering accuracy derives directly from data in the standard which must be followed without exception.

Sizing

Sizing an orifice depends on the flow measurement task to be done. For example, a simple design would be a single meter tube with a single orifice plate in the mid beta ratio range, which would be sufficient for a fairly constant flow. However, if the flow grows over time, then sizing would allow for this growth by using a 0.20 beta plate. If flow is likely to decrease with time, then a large beta (such as 0.60) should be used. This way the meter tube size will not have to be changed in a short time period. If continued growth is anticipated, an economical design would be to size the lengths and valves for the meter tube size slightly larger (i.e., if a 6 inch tube is chosen, size the meter tube length for an 8 inch tube).

On non-custody transfer metering where utmost accuracy is not of prime importance, design betas can be changed to 0.15 and 0.70. If flow changes take place in a short time period, consideration should be given to several ranges for differential devices tied to one orifice or the use of multiple meter tubes switched in and out of service automatically. The differential chosen to size an orifice should not be full scale, but a reduced differential to allow for some inevitable variations. The amount of this reduction might be 10 to 20% of range depending on the likelihood of this much variation occurring.

Sizing of orifice stations is relatively simple for fluids with steady flow rates. Since a single orifice with a single readout system is limited to a flow range of roughly 3 to 1, for most accurate measurement, a knowledge of the flow ranges is required to properly size the orifice to prevent over- or under-ranging. The controlling design flow rate should be the "normal flow" since the majority of the flow will, by definition, be at this rate. If the limits are beyond the 3-to-1 range, then multiple orifice or readouts or possibly another meter should be chosen. There are many calculator and computer programs available in the industry to assist in this sizing. Likewise, manufacturers offer sizing calculations as part of their service of manufacturing plates.

Equations

The basic orifice meter mass flow equation in the common U.S. system of units (IP) is as follows:

$$q_m = C_d E_v Y(\frac{\pi}{4}) d^2 [2 g_c \rho_{t,p} \Delta P]^{0.5} \quad (20)$$

Where:

q_m = mass flow rate (lbm/sec)
C_d = orifice plate coefficient of discharge. (No units)
d = orifice plate bore diameter calculated at flowing temperature (T_f). (feet)
ΔP = orifice differential pressure (lbf/in^2)
E_v = velocity of approach factor (no units)
g_c = dimensional conversion constant ($lbm\text{-}ft/lbf\text{-}sec^2$)
π = universal constant (3.14159)
$\rho_{t,p}$ = density of the fluid at flowing conditions (P_f,T_f), (lbm/ft^3)
Y = expansion factor (No units)

In this equation q_m is the mass flow rate and is the value to be determined. C_d is the coefficient of discharge that has been empirically determined; it has been changed in the 1991 edition of the ANSI/API 2530[5] (AGA-3)[1] documents. It depends on the plate and meter run size. E_v, the velocity of approach factor, corrects for the change in the flow constriction shape for with various beta-ratio plates — as well as differential pressure and static pressure effects — as the flow goes from a meter run to the orifice plate restriction. Y, the expansion factor, corrects for the change in density from the pressure tap to the orifice bore in gas measurement. It can be calculated from either an upstream or downstream tap depending on which tap measures the static pressure. Its value is a function of the beta ratio, the differential pressure, static pressure at the designated tap and the isentropic exponent of the flowing gas. The factor is unity for liquid flows. $\pi/4$ is used to convert the orifice diameter to an area. d is the orifice plate diameter as determined by proper micrometer readings. g_c is the attraction of gravity constant assumed at base conditions. $\rho_{t,p}$ is the density of the fluid at the flowing temperature and flowing pressure. ΔP is the differential pressure measured across the orifice. All values must be in a consistent set of units, as shown.

Most standards reduce the basic mass flow equation (Equation 20), into one that allows the more convenient use of mixed units and reduces constants to a single number. However, all versions trace back to this general basic equation (Equation 20).

Equation 20 can be converted to volume at base conditions by dividing the equation by the fluid density at base conditions:

$$Q_v = \frac{q_m}{\rho_b} \tag{21}$$

Since measurement is often done with mixed units for convenience, the simplified equations contain a numerical constant for balancing the units. For example, in natural gas measurements, Equation 21 may be changed to:

$$Q_v = 218.527 C_d E_v Y_1 (d^2) [\frac{T_b}{P_b}] \left(\frac{P_{f1} Z_b h_w}{G_r Z_{f1} T_f} \right)^{0.5} \tag{22}$$

where: Q_v = standard volume flow rate - SCF/hr.
218.527 = numerical constant and unit conversion factor
C_d = orifice plate coefficient of discharge
d = orifice plate bore diameter calculated at flowing temperature (T_f) - in.
G_r = real gas relative density (specific gravity)
h_w = orifice differential pressure in inches of water at 60 °F
E_v = velocity of approach factor
P_b = base pressure - psia
P_{f1} = flowing pressure (upstream tap) - psia
T_b = base temperature - °R
T_f = flowing temperature - °R
Y_1 = expansion factor (upstream tap)
Z_b = compressibility at base conditions (P_b, T_b)
Z_{f1} = compressibility (upstream flowing conditions - P_{f1}, T_f)

Similar simplifications are also made in equations so applications are simpler for standard units of the measured variables for liquid flows.

For these equations to be valid with minimum errors, the following factors are vital:

- The orifice plate and meter run must be kept clean and retain the original conditions specified by the standard.
- To assure this, periodic inspection should be conducted to reaffirm conditions. Inspection frequency depends on problems of foreign material collection and possible damage. Inspections will confirm the orifice diameter and coefficient of discharge.

The expansion factor (Y_1) for gas provides a relatively small correction for high pressure (above 100 psia) measurement provided the differential, in inches of water, is not allowed to exceed the static pressure (psia). As this ratio increases(below 100 psia), the correction factor increases, as does the error in the factor.

Flowing density can be measured directly with a densitometer that is periodically calibrated. It also may be calculated from appropriate equations of state whose accuracy must be established from the data upon which the equation is based. In either case, density is usually the second most important variable to determine in the equation.

The most important variable in the equation — the one that directly determines flow-measurement accuracy — is the differential pressure. Therefore, a major effort should be made to have as high a differential as possible (considering flowing conditions), and the best available differential transducers should be specified. Anything that can be done should be done to improve this most critical measurement factor.

Maintenance

Orifice meter maintenance consists of periodic inspection (as indicated above), cleaning primary elements, and scheduled testing and calibration against standards (if necessary) of the secondary elements. Maintenance frequency, if not set by agreement or contract, should simply be based on experience and performed as often as necessary to correct any calibration drift or error that may occur. Proper records for each station will determine this schedule.

Advantages of the orifice meter:

1. Well-documented in standards
2. Enjoys wide acceptance; personnel knowledgeable across the industry about requirements for use and maintenance
3. Relatively low cost to purchase and install
4. No moving parts in the flow stream
5. When built to standards requirements, does not require

calibration beyond confirming mechanical tolerances when purchased and periodically in use
6. Rugged, stands abusive flow conditions well; inexpensive to replace if damaged
7. No limitations of temperature, pressure, or corrosion with choice of proper materials
8. Electronic readout systems available to provide immediate calculation of flow rate and total flow; no manual errors or prolonged time delay to get data as is the case with charts.

Disadvantages of the orifice meter:
1. Low rangeability with a single readout
2. Relatively high pressure loss for a given flow rate, particularly at lower beta ratios
3. More sensitive to flow disturbances at higher beta ratios than some meters
4. Flow pattern in the meter does not make meter self-cleaning.

Special-shaped orifices

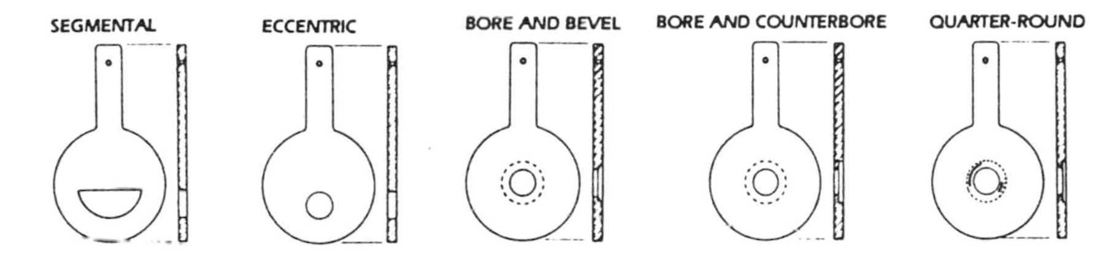

Figure 8-7 Many variations in orifice design allow for special measurement applications.

Several special-shaped orifices are worthy of mention. The quadrant-edge (quarter circle) orifice is used at Reynolds numbers below which the square-edge concentric orifice coefficient becomes too nonlinear to be useful. A conic-entrance orifice can be used for a similar range of Reynolds number; however, it can be applied for even lower than quadrant-eged Reynolds numbers. Both of these devices should be used only with limited diameter ratios, and the required flow rates and Reynolds numbers must be evaluated carefully to assure good measurement. (See Chapter 6, Reference 5.)

Eccentric, segmental and annular orifices — with accuracies in the order of 2% — are special devices to take care of dirty fluids and two-phase flows. Since these are not the best devices for obtaining accurate measurement, they are used only where these special conditions exist. No detailed standards exist for these devices. For details on construction of these orifices. (See Chapter 6, Reference 5.)

Honed flow sections are orifice runs made in sizes of 0.25 through 1.50 inch. They are covered by the ASME (Reference 2 at the chapter end), and data are available from manufacturers who have developed special manufacturing requirements and special coefficients to calculate corrected flow. These devices are used for low flows of gas, liquid, and steam with a higher tolerance than standard sized meter tubes of 2.00 inches and larger.

Flow Nozzles

Another important flow element is the flow nozzle. Several configurations are available, the most important of which are the ASME[2,7] long radius nozzle, high or low beta series, and the throat tap nozzle for gas and liquid. In Europe, another nozzle shape outlined in ISA-1932 is more often used than the ASME nozzle. Both have the same rangeability limitations as orifices: approximately 3-to-1 for a 100-inch recorder. (See ASME MFC-3M[7] and ASME PTC-6[6] for details of construction.)

When flow rates change with time and thus require changing nozzles, it is more difficult than changing an orifice plate. The nozzle, however, is better able to sweep suspended particles by the restriction because its contour is more streamlined than the orifice. The ability to handle particulates is particularly good if a nozzle is installed in the vertical position with flow downward.

Nozzles are used mostly for high-velocity, nonviscous, erosive fluid flows. However, they have considerable acceptance in certain industries such as electrical power generation. Standard nozzles are moderately priced, but throat tap nozzles are very expensive. The throat tap nozzles provide some of the highest accuracies of all primary devices, since they are allowed in only a very limited beta ratio range of 0.45 to 0.55.

The special throat tap type is used primarily for accurate acceptance testing of electrical generating plants. This is a very expensive nozzle because it must be flow calibrated prior to use, and its calibration must meet the standard coefficient value within ± 0.25%. This doesn't allow much room for error in manufacture or calibration. Because of these problems, its use is primarily restricted to the power industry where the cost of acceptance testing can justify the nozzle cost.

The devices are very difficult to remove for inspection and cleaning, and their use in fluids where deposits may build up is not recommended. Installation requirements for nozzles are similar to those for orifices; requirements are detailed in the ASME[2,7] documents previously listed.

The nozzle has mistakenly been said to have low pressure drop. But for a given differential and pipe size, it is better stated that a lower beta ratio can be used. At times, a smaller tube size can be used with the nozzle than with the orifice. Permanent pressure drops will be approximately equal for a given set of flow conditions for either device if the differential range is set. Nozzle sizing must be based on good flow data that is fairly stable because of its restricted range. The expense of incorrect sizing should always be kept in mind.

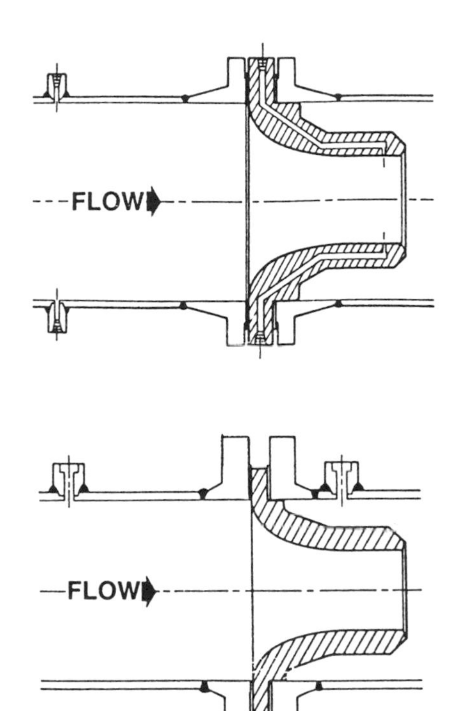

Figure 8-8 Flow nozzles — the top unit shows a throat tap, and the bottom unit represents an ASME "low-beta" design.

ASME[2,7] nozzle taps are located in the pipe wall one pipe diameter upstream and one-half pipe diameter downstream. The ISA-1932 nozzle uses corner taps. Nozzles are usually mounted between pipe flanges.

Nozzles may be fitted with a diffuser cone to reduce the pressure loss by guiding flow back to the meter tube. It is possible to truncate this diffuser to approximately 4 times nozzle throat diameter and have about as full recovery as with a cone extending to the wall.

The equation for the ASME[2,7] (low or high beta) flow nozzle is in the same form as the orifice, with the ASME nozzle coefficient restricted to a Reynolds number of 10,000 or greater. The coefficient value is approximately 0.95 versus an orifice coefficient of approximately 0.60, so the nozzle handles over 50% more flow for a given size and differential reading. A special equation can be used for the coefficient at lower Reynolds numbers, but the change of coefficient in this range requires an iterative process of flow velocity and Reynolds number to maintain accuracy. Nozzles with other shapes have different coefficient equations.

Operational accuracy of the nozzle is again directly related to the ability to measure at higher differentials because of the square-root relationship with flow. It is best not to measure below 10 inches of differential, but higher differentials (400 to 800 inches of water, for example) can be used because of the strength of the nozzle versus the orifice.

The nozzle should be applied to clean fluids since removal for cleaning is very difficult. Any change of nozzle throat finish has a direct effect on measurement accuracy. Where critical measurements are made, the installation must be capable of being shut down, or have a bypass to allow periodic inspection and cleaning. This increases the cost of a nozzle installation. For the most part, nozzles have very little maintenance done on them.

Advantages of flow nozzles:

1. Can be used at higher velocities with little damage to the nozzle surface, which allows using smaller sized meters for a given flow rate
2. Throat tap nozzles have the lowest stated tolerances of all head devices, but require calibration to prove and have restricted beta ratio ranges

Disadvantages of flow nozzles:

1. Expensive
2. Difficult to install and remove for cleaning — and therefore are seldom cleaned
3. Data base for coefficient tests much smaller than for orifices; calibration required for best accuracy
4. Limited range of Reynolds numbers

Venturi Meters

Low-pressure-drop devices are headed in the U.S. by the classical Venturi which is used for liquid, gas and steam at pipe Reynolds numbers above 100,000. Venturi rangeability is likewise dependent on the differential readout and is 3-to-1 for a single 100-inch range recorder. Venturis are mainly used to measure liquids, clean or dirty.

The Dall tube is popular in the U.K. as a low pressure device for these same measurements.

Where low pressure drops are required on nonviscous fluids, the amount of pressure drop is dependent on the angle of the downstream cone and the beta ratio of the Venturi. Venturis have the disadvantage that their size makes inspection and changing cumbersome. It is recommended that an initial calibration be run for precision measurement since they tend to be more difficult to manufacture precisely than a nozzle or an orifice. These problems can be secondary to savings achieved in the costs of low operating pressure drops.

The classical (Herschel) Venturi is built to specifications of the ASME MFC-3M[7] and ISO-5167[2] with an angled inlet and exit cone on a cylindrical throat section. It is designed to minimize pressure drop and to be somewhat self-cleaning since there are no sharp corners in the flowing stream where materials may collect. Venturis have an overall length of approximately 8 diameters; this makes them unwieldy to make, install and remove for inspection. The actual design has several options that must be chosen before the length and shape dimensions are set. One sometimes overlooked is inlet cone finish, which affects coefficient limits. Each Venturi is unique, and consequently the volumes to be measured must be accurately known because once a Venturi is built, there is no flexibility in the unit.

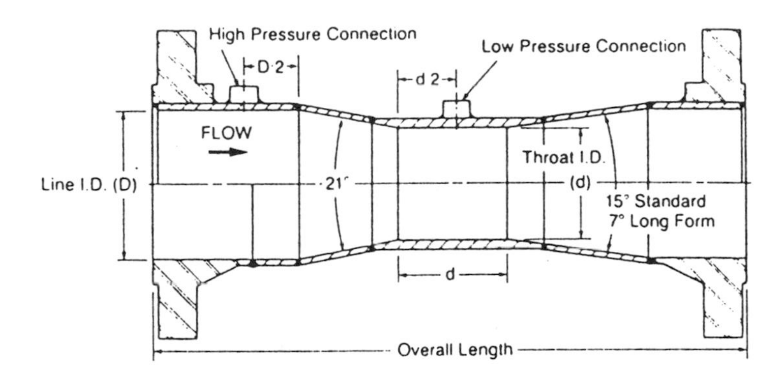

Figure 8-9 Herschel Venturi.

Venturi installation

Installation requirements for piping depend upon the upstream fittings (i.e., elbows, valves, reducers, and expanders) but are generally shorter than for a nozzle or orifice with the same beta ratio. No straightening vanes are required by the standards. However, experience shows that no swirl can be tolerated coming into a Venturi, so two elbows in different planes or pinched valves should not be installed in upstream piping.

In summary, Venturis are 3-to-1 flow range devices with a single 100-inch manometer recorder. Their throughput has the highest efficiency of any of the head devices with an approximate coefficient of 0.98. They should be designed to the actual conditions of flow with no design allowance for errors, thus the flow specifications must be correct.

The flow equation is the same as for the orifice previously covered except for the coefficient of discharge. A Venturi should perform correctly over time provided the surface of the inlet cone and the throat are not changed by corrosion, erosion or deposits. Venturis can be cleaned if provisions are made to remove the unit from service for maintenance. Other than this (which is not often done), most maintenance consists of work with the secondary equipment.

Other low loss devices are the Universal Venturi and the LoLoss tube.

Advantages of Venturis (and other low-loss head devices):

1. Low permanent pressure loss
2. Can be used on slurries and dirty fluids

Disadvantages

1. Limited rangeability; must be used only on installations where the flow rate is well known and varies less than 3-to-1
2. Very expensive; should be flow calibrated to provide accuracy into the range of ±1.00%
3. Units are big and weigh more than comparable head devices; this makes them difficult to install and inspect.

Turbine meters

Turbine meters are used successfully and widely in both liquid and gas measurement. They are made differently for gas and liquid measurement because of the difference in driving forces of the fluids and internal bearing frictions. However, the basic operation is the same for gas or liquid service.

The turbine meter is a velocity measuring device. Flow passes through a free-turning rotor mounted coaxially on the meter body centerline and exits the body. Since velocity is the parameter measured, the upstream and downstream piping have defined lengths to eliminate nonstandard velocity profiles and swirl.

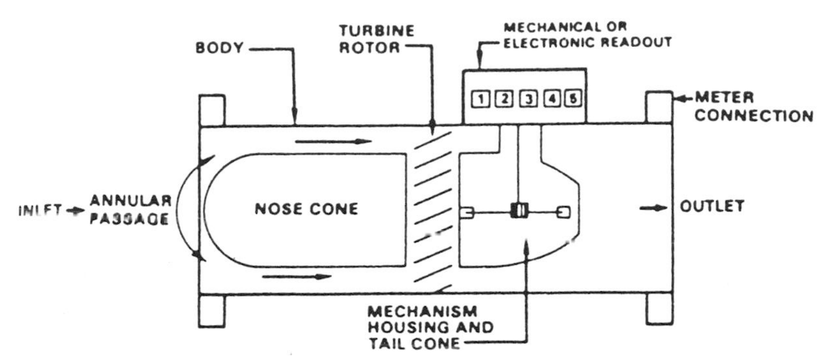

Figure 8-10 *Schematic drawing of an axial flow gas turbine meter.*

Fluid imparts an angular velocity to the angled rotor so that the rotation is proportional to flow rate. Blade shape and angle, bearings style, and other construction details vary from manufacturer to manufacturer. With accurate measurement of rotor speed from mechanical gearing or magnetic pickup and by knowing the hydraulic area that the flow is passing through, volume at line conditions can be determined. Since no rotor is without friction, the aerodynamic friction across the rotor and the friction of the bearing system cause a nonlinear calibration until the retarding forces are a small percentage of the driving forces. At this point, the calibration curve becomes linear (rotor speed increases directly with flow velocity).

Figure 8-11 *Typical liquid turbine meter. This design uses an internal design such that upstream and downstream forces on the rotor balance for a "floating" rotor.*

Actual flow area is not the calculated open area, but something less because turbulence blocks some of the area. Because of the friction, turbulence, and necessary manufacturing tolerances, each meter must be calibrated to determine its proof curve. The necessity for this calibration is different than for an orifice which can be manufactured to a tolerance that allows its calibration to be predicted from its mechanical shape; calibration for a turbine meter cannot be determined this way.

As flow through the turbine first increases from zero, a certain amount of fluid passes through the rotor before it begins to turn. At some point, the fluid imparts enough force to overcome the friction retarding forces of the rotor bearing. At this point, the rotor begins to turn and the frictional forces in the bearing become small. The aerodynamic forces predominate and control the rotor's speed. The existence of these retarding forces and the slight change in flow area create a difference between the theoretical and actual rotor speed. These differences must be accounted for with a calibration run on each meter. As flow rate increases, these aerodynamic and bearing friction forces become minimal and the proof curve becomes linear reflecting only an increase in velocity.

There is another kinetic effect to consider. Fluid entering the meter is speeded up by a deflector before it passes through the rotor. More driving force results on the rotor because of the increased velocity and the fact that the average velocity is being imparted further out on the rotor. This improves the performance curve at lower flow rates. The flow deflector also serves to lessen thrust loads on the rotor bearing by shielding the center of the rotor from the flowing stream.

For gas meters, the deflector is larger (the annular opening smaller) so it blocks approximately 66% of the meter area versus less than 20% for liquid meters. This is because of the need to generate higher velocities, hence torque, on the gas meters that operate in fluids with densities lower than existing with liquid meters. The gas meters use sealed or shielded ball bearings for minimum friction; liquid turbines use sleeve bearings for rugged wear characteristics.

Rangeability

Rangeability of a gas turbine meter varies with pressure — from approximately 10-to-1 on gas at atmospheric pressure to over 100-to-1 on gas with pressures over 1000 psia. On the other hand, liquid meters maintain a constant range of approximately 10-to-1 but have some overriding concerns for changes in viscosity, density and meter size. As the viscosity goes above that for water, the meter range can be diminished down to as low as 3-to-1. Likewise, as densities drop to 60% of the density of water and lower, the range begins to decrease until it may reach 3-to-1. Smaller sizes (below 6 inch) tend to have lower ranges of linear proof curves. The specific manufacturer should be consulted as to degradation of a liquid meter with viscosity, density and size.

Meter capacity is determined by allowable rotor speed (bearing speed limit), pressure drop, and fluid velocity (blade angle). All manufacturers choose different design parameters so their specific meter sizes handle given volumes which may be similar but not equal to another manufacturer's.

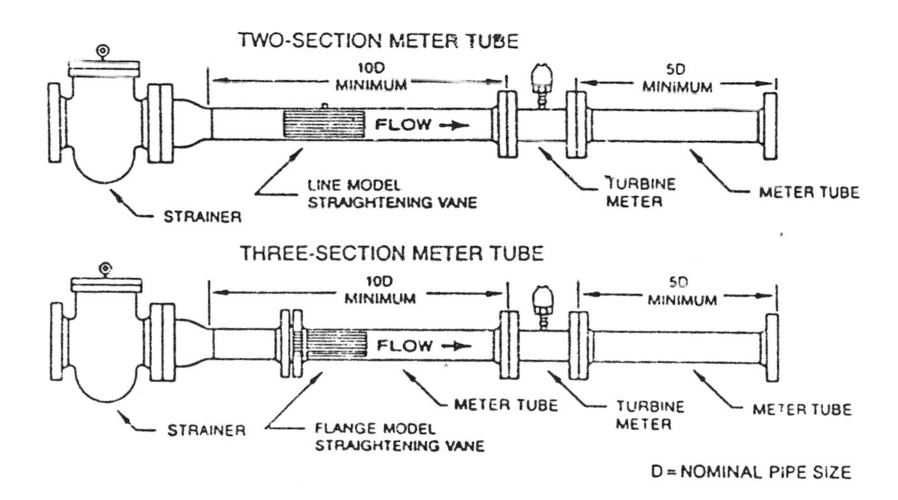

Figure 8-12 *Installation of two- and three-section gas turbine meters according to the AGA-7 requirements.*

Installation

Installation of a gas turbine must be according to AGA-7[4] or ISO-9951[9] (Draft). Liquid turbine meters must be installed according to Chapter 5.3 of the API Manual of Petroleum Measurement Standards[3] for the custody transfer type of metering. Other standards are more lenient in required lengths. Piping to allow for testing and removal of the meter for repairs is necessary since neither liquid nor gas meters can be removed from an operating line without stopping flow and depressuring. If the delivery requires flow continuity, then a bypass must be installed. Note: The laws of some countries require metering continuity on any custody transfer meter. If this is the case, then the bypass must include a meter. Sizing tables supplied by the meter manufacturer should be used in designing a meter station. If flow rates fluctuate, the range of the flows should be maintained within the turbine meter limits, particularly for low flow rates where the meters have larger errors if limits are exceeded.

Equations

Equations for the gas and the liquid meters are different. The gas turbine meter equation is as follows:

$$q_b = q_f M_f \left(\frac{P_f}{P_b}\right)\left(\frac{T_b}{T_f}\right)\left(\frac{Z_b}{Z_f}\right) \tag{23}$$

where: q_b = Flow Rate at base conditions
q_f = Flow rate at operating conditions (meter reading)
M_f = Meter factor to correct meter output based on calibration
p_f = Pressure flowing conditions
p_b = Base pressure set by agreement near atmospheric pressure
T_b = Base temperature set by agreement near atmospheric temperature (60°F)
T_f = Temperature at flowing conditions
Z_b = Compressibility at base pressure and temperature
Z_f = Compressibility at flowing pressure and temperature.

The liquid turbine meter equation is as follows:

$$q_b = q_f M_f F_t F_p \tag{24}$$

where: q_b = Flow rate at base conditions
q_f = Flow rate at operating conditions (meter read'g)
M_f = Meter factor to correct output based on calibration
F_t = Factor to correct fluid from flowing temperature to base temperature
F_p = Factor to correct fluid from flowing pressure to base pressure.

A turbine meter operates over its specified range with equal accuracy. Overrangeing by pressure drop can damage blades, or high velocity can damage bearings. This is especially a problem while putting meters in and taking meters our of service; at these times, flow rates must be changed slowly. Meters used with liquids that vaporize as pressure is removed from them require special filling techniques so that the meters are not damaged. This can be done by slowly filling the system with a gas while monitoring the rotor speed. Continue filling until the pressure of the liquid is reached. The gas can be bled off slowly while the liquid is allowed to displace it without vaporization. When the liquid has completely filled the system, liquid flow may be started.

Maintenance and calibration

Maintenance for properly operated turbines consists of periodic cleaning and physical inspection. Calibrations may be required to reconfirm proof curves on custody transfer meters. This may be by calibration against standardized master meters or direct calibration against standards (i.e. critical flow nozzles for gas, pipe provers for liquids).

Advantages of turbine meters:

1. Good accuracy over full linear range of meter (accuracy is per cent of flow rate, not percent of full scale)
2. Electronic output available directly at high resolution rate which makes proving possible in a short time period
3. Meter cost is medium, but total meter station is low-to-medium cost because of high flow rate for given line size
4. Has pressure and temperature limits, but can handle normal flow conditions very well
5. Excellent rangeability on gas meters at high pressure

Disadvantages of turbine meters:

1. Require throughput proving to establish most accurate use
2. Viscosity effects liquid meters that may require separate proof curves for different viscosities
3. Rangeability at low pressures about the same as other gas meters
4. Require upstream flow pattern to be non-swirling, which necessitates straightening vanes or very long inlet pipe.

Positive displacement meters

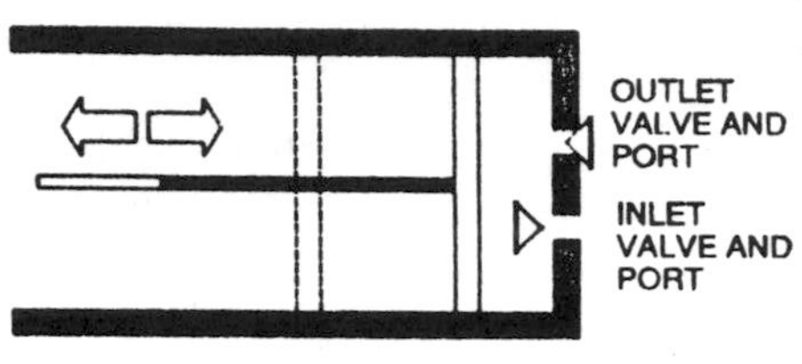

Figure 8-13 Cylinder and piston show principle of displacement metering.

Positive displacement (PD) meters are used for measurement of liquids and gas. Two of the most common meters at residences are the water and gas meter, both of which are usually positive displacement meters.

The basic PD design is a "bucket" alternately filled and emptied. To keep operation from being a strictly batch operation, most PD's have multiple "buckets" that are geared and valved together so that while some buckets are filling, others are emptying. With proper timing and valving, there is an uninterrupted flow through the meter. The driving force for this action comes from the flowing stream as a pressure drop.

When stopped, most PD's create a large blockage factor in the line. If there is danger of the meter being jammed, some means of bypass or relief must be designed, otherwise flow may be reduced to a very small percent of normal flow. In some cases a severe flow reduction may not be a critical concern, unless the pressure drop builds to a point that major damage is done to the meter. This problem is somewhat unique to PD's and must be examined before metering system designs are set.

Rangeability

One characteristic of PDs that make them attractive is rangeability unmatched by other meters. Likewise, they are able to measure very small flows in their stated ranges with relatively good accuracy. Since the PD meter's flow path is normally not straight through, requirements of upstream and downstream piping are minimal and are normally based on piping and space concerns rather than flow pattern considerations. Periodic maintenance and calibration must be allowed for in the design with provisions to interrupt or bypass flow.

Design

Design of PD meters vary from device to device and serve various portions of industry needs depending upon operational characteristics. For example, common to the natural gas industry are four-chamber/two-diaphragm meters, three-chamber meters with an oscillating valve, rotary meters of the roots

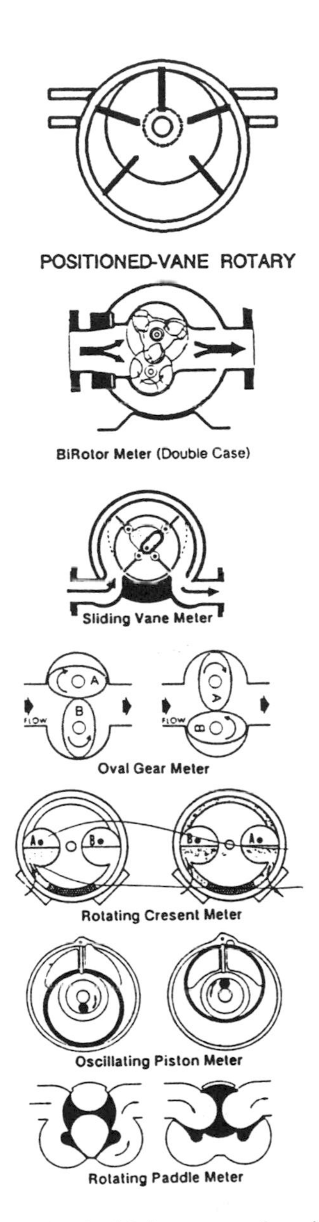

Figure 8-14 *Major types of positive displacement (PD) meters for liquid measurement.*

type, and meters with four rotating vanes and rotoseal meters. Liquid PD types include reciprocating piston, ring piston, rotating discs, sliding vane, rotating vane, oval gear, nutating disc, metering pumps, and lobed impeller. Standards and manufacturers' literature should be consulted for operational details, sizing, range and accuracy limits, and recommended applications.

Because of the mechanical size of PD meters, limits on pressure, cost and weights become increasingly important as the meters get larger than about 10 inch. They are made in larger sizes, but the applications are generally limited to special flow measuring situations.

Performance

PD meters perform well for long periods of time provided fluids are clean, non-erosive, non-corrosive, non-depositing, and proper maintenance is routinely performed. In less critical measuring systems, such as domestic water and gas, meters are run for many years with testing only on the basis of a statistical failure-rate study or upon customer complaint.

On the other hand, large liquid PD meters used in the petroleum industry may be tested on a daily or weekly basis with a prover system permanently installed as part of the metering station.

There should be no over-ranging; if necessary, a protective flow limiting device (automatic bypass valve) should be installed to prevent the over-ranges.

Equations

The equations for the PD meters are the same as for turbine meters; readout of the PD, which is at line volume conditions, is multiplied by a meter factor arrived at from proving and correcting factors to reduce the flowing conditions of pressure and temperature to the base or contract conditions. (See Equations 23 and 24.)

Maintenance

Maintenance of small PD meters (below 4") usually involves unit replacement with repairs or rebuilding done at a central meter shop rather than in the field. With larger meters (6" and above), maintenance consists of part(s) replacement in the field and removal to a meter shop only if field repairs are unsuccessful. Maintenance decisions are based on the economics of maintenance costs versus the cost of inaccurate measurement.

Advantages of PD meters:

1. Insensitive to upstream and downstream piping effects so that no or minimum lengths are required
2. Operating principle straightforward, easy to understand
3. Rangeability highest of liquid and gas meters available without loss of accuracy
4. Even though valving and clearances require close tolerances, commercially available units are rugged and provide long and reliable service on clean fluids or with line filters
5. Simple to complex readout systems available for a simple flow equation

Disadvantages of PD meters:

1. Because of clearances required, pressure, temperature, and viscosity ranges are limited and special care may be required for installation
2. For larger sizes (above 10"), meters are large, heavy, and relatively expensive
3. Head loss can be high, particularly if the meter jams — protection from flow shutdown and pressure overrange may be required

4. Filtration or strainers may be required for fluids containing foreign particles to minimize meter wear
5. Maintenance costs high on larger meters; unit replacement typical for smaller meters because of complexity and cost of repairs.

Vortex shedding meters

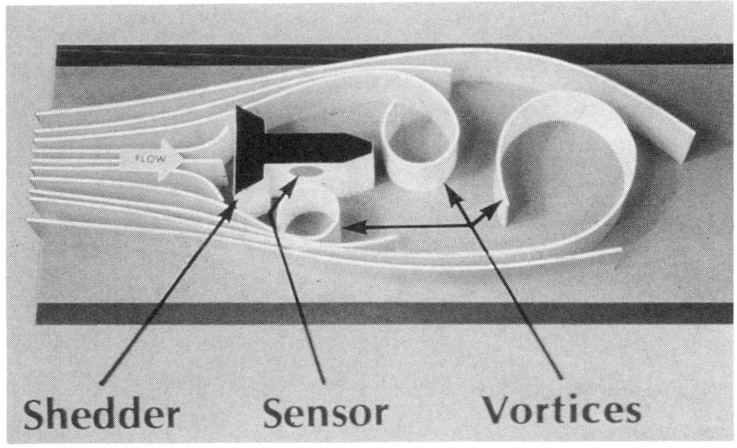

Figure 8-15 Major components in a vortex-shedding meter.

The Vortex Shedding meter has come into prominence and usage in the last 10 years for both gas and liquid measurement. It has received acceptance in the industrial flow measurement area and, to a limited degree, the custody transfer measurement area. Although based on the same basic principle, vortex shedding meters are standardized by performance, not in terms of mechanical construction of primary and secondary elements of the meter.

Operation

The vortex shedding meter operates on the Von Karman effect of flow across a bluff body. This principle states that flow will alternately shed vortices from one side and then the other of a bluff body, and the frequency of shedding is proportional to velocity across the body. When this velocity is combined with the hydraulic area of flow in a stream, the rate of flow can be established. Action is similar to the movement of a flag downstream of a flag pole. The rippling of the flag is due to the vortices as they are shed alternately on each side of the flag pole. Counting the vortices can be done in many ways since the vortex represents a pressure and temperature change, and either of these may be sensed. Or, a secondary effect of the small movement at the bluff body can be sensed.

In any case, vortices are shed irregularly at low flows. When these stabilize, the meter's lower flow rate is defined. Manufacturers have made continual developments in the readout systems and in determining the best bluff body shape to give a strong stable shedding pattern. Because the individual differences in the bluff body and the readout, each design is unique, and meter calibration should be obtained from the manufacturer.

Since the meter reacts to velocity, it follows that a proper flow pattern must be presented to the bluff body. This is accomplished by using straightening vanes and/or straight upstream piping to eliminate swirl and distorted patterns. Installation requirements are similar to other velocity sensitive meters.

Sizing

The sizing of these meters with normal pipeline velocities make throughput per line size higher than many other meters. Manufacturers' sizing recommendations should be followed.

Equations

The equations for vortex flow meters are the same as those for turbine meters (Equations 23 and 24) since the meter produces a pulse output proportional to the flow rate at line conditions, and this output must be corrected from line conditions to base conditions. Depending on the individual meter, a calibration factor K is determined which relates produced pulses to the line volume passed versus Reynolds number. These factors are supplied by the manufacturer based on calibrations covering a range of Reynolds numbers (liquid and gas) similar to the operating Reynolds number. The K curves are quite linear for flows above the low-end limit. Viscous liquid should be checked to make sure its Reynolds number at flowing conditions will exceed the low-flow limit, usually in the 10,000 range.

Maintenance

As long as the bluff body and the meter body opening are kept clean, the meter should retain its original calibration. Any erosion, corrosion, or deposits that change the shape of the bluff body will cause a change in the hydraulic area and shift calibration. Periodic inspection is recommended to ensure that initial conditions are being maintained in the primary element.

The flow-variable correction instruments of the secondary system must be calibrated to insure that the transducers have not changed calibration. If recalibration of the primary element is required, then some type of throughput test is run against a standard.

Advantages of vortex-shedding meters:

1. Relatively wide rangeability with linear output
2. On clean fluids (liquids and gases), the meters have long-term stable proofs
3. Frequency output can be read directly into electronic readout systems
4. Installation costs moderate, installation simple
5. When minimum or higher Reynolds numbers pertain, effects of viscosity, pressure and temperature are minimal
6. No moving parts in contact with the flowing stream

Disadvantages of vortex-shedding meters:

1. Flow into a meter must be swirl-free; this requires straightening vane and/or long, straight piping
2. Output may have "jitter" (frequency instability) and/or fade in certain areas of operation which affect readout requirements
3. Not available in sizes above 8 inches
4. Pulse train is irregular, proving requires a long test time to obtain a representative average pulse rate
5. Pulse resolution the same for all meter sizes; this means a low pulse rate with larger meters yields low volume resolution
6. Subject to range limitations at lower Reynolds numbers.

Ultrasonic meters

The ultrasonic meter category contains a number of different designs for measuring an average velocity in a flowing system. They are all based on an ultrasonic signal being changed by or reflected from the flowing stream velocity. Meter accuracy relates to the ability of the system to represent the average velocity over the whole stream passing through the meter body hydraulic area. This ability affects installation requirements and accuracy of results obtained.

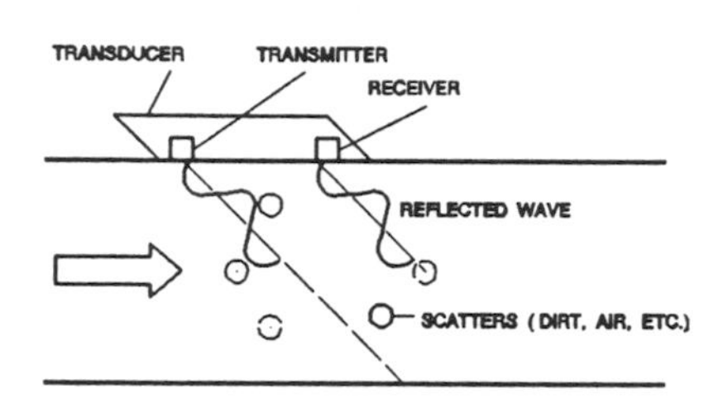

Figure 8-16 Doppler reflection meter.

Dopplers

The two main types of ultrasonic meters are the Doppler frequency shift and the transit-time change. The Doppler meter is used on liquids and gases with some type of entrained particles that are traveling at the same speed as the main body of flow. The ultrasonic signal is reflected from these traveling particles across the stream, and the shift in the frequency is related to the average velocity of these particles over time. Meters are made in several types; one type requires installation of transducers into the flowing stream, the other is a strap-on model that can be installed without shutting down the flowing stream.

Transit-time

A transit time unit is installed directly into the flowing stream and can be made with single or multiple transducers for establishing average velocity. These units can be used on liquids or gas. The multiple transducer units can handle velocity profile distortions (including swirl) so that installation requirements are reduced. But meter complexity (i.e., cost) goes up because of the multiple transducer units and the more complex electronics required to compute average velocity and flow.

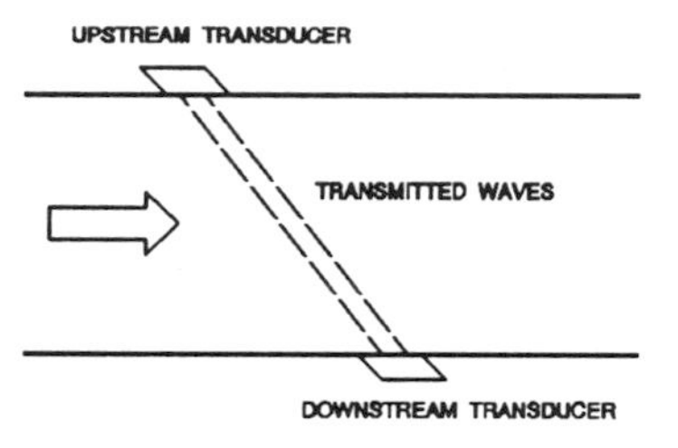

Figure 8-17 Transit-time ultrasonic meter.

The multipath meter uses transducers set at an angle to the flow axis. Each transducer in a pair functions alternately as transmitter and receiver over the same path length. When equations for transit times "upstream" and "downstream" are used to determine mean transit time, the speed of sound in the medium drops out. Consequently, gas velocity through the meter can be determined from transit times and physical dimensions of the spoolpiece.

Figure 8-18 Transducers in this multipath ultrasonic meter are mounted so the sound travel path crosses the pipe body at an angle. Four pairs of transducers are used for high accuracy.

Equations

Volume flow rate equals weighted, calculated mean velocity times meter-bore cross-sectional area. To convert from flowing volume at base conditions, corrections must be made for pressure and temperature as for a turbine meter. Here are the equations involved for a transit time meter:

$$t_{ud} = \frac{L}{C+V_m\cos\Theta} \quad (25)$$

$$t_{du} = \frac{L}{C-V_m\cos\Theta} \quad (26)$$

$$V_m = \frac{L}{2\cos\Theta} x \frac{t_{du}-t_{ud}}{t_{du} x t_{ud}} \quad (27)$$

Actual volume flow rate (Q) is:

$$Q = V_m x \frac{\pi D^2}{4} \tag{28}$$

$$Q = K \frac{t_{du} - t_{ud}}{(t_{du})(t_{ud})} \tag{29}$$

where: t_{ud} = transit time from transducer U to D
t_{du} = transit time from transducer D to U
L = distance between transducers U and D (path length)
C = velocity of sound in the gas
V_m = mean velocity of the flowing gas
θ = angle between acoustic path and meter axis
D = diameter of the meter bore
K = constant for a specific meter

Performance

An ultrasonic meter's performance depends on its ability to find the average velocity, the condition of the meter open area (no change with flow rate), and abilities of the readout system. Meter calibration, based on transit time, relates directly to the mechanics of construction, as discussed above. It can be calculated and checked by a mechanical inspection to determine geometrical dimensions. It can also be checked by filling the meter (during no flow conditions) with a fluid whose speed of sound is known (such as nitrogen) and calculating the transit time over the signal path. Calibrations can also be run against volume standards or master meters.

Maintenance

There are no moving parts that require lubrication, and maintenance is basically performed only on the readout system for meters installed to measure clean fluids. For dirtier fluids, cleaning must be done if the meter's or transducers' flow areas are affected. The meter causes no pressure drop other than the normal drop in an equivalent length of pipe. The meters have a large turndown ratio with accuracies for multi path designs equal to the best of other types of meters; single path designs are more sensitive to flow pattern irregularities and provide less accuracy. Data are available on the accuracies of these meters from independent laboratories, users, and manufacturers. The ultrasonic principle is applicable to all pipeline sizes; manufacturers' literature lists available meter sizes. Bidirectional flows are measured with no additional electronics, mechanics or piping configuration since the change of timing reflects the flow reversal, and the electronics calculate flow accordingly. Accuracy is the same in either direction as long as the minimal upstream requirements are met by piping on both sides of the meter. These requirements change depending on the type of ultrasonic meter used. The meter is available in wide temperature and pressure ranges.

Advantages of ultrasonic meters:

1. No pressure drop, since meters are same diameter as adjacent piping
2. High frequency pulse rate of output minimizes errors from effects of pulsation and fluctuating flow
3. Installation can be simple and inexpensive
4. High rangeability
5. No moving parts in contact with flowing fluid
6. Simple mechanical calibration easily checked without a throughput test

Disadvantages of ultrasonic meters:

1. Power required for operation
2. Flow profile must be fully developed for an average velocity to be determined from a single path or reflection unit. (Note: multiple-path units average disturbed flow patterns including swirl to minimize flow profile problems).
3. High initial cost

Other meters

Pitot tubes (standard and averaging) can be covered in a single section since they are similar in their application limitations. They are used for the measurement of liquids and gas. The meters do not "look" at the total flow profile, and their accuracies relate to how well the readings represent an average velocity. When a single point in the line velocity is measured, accuracy generally ranges to ± 5% unless specific calibrations are run.

Over limited ranges, these devices are used for control purposes rather than custody transfer. Their major advantages are that they create practically no pressure drop and are inexpensive. On the other hand, velocities in normal pipeline designs are such that measured differentials in Pitots (which may be a maximum of 8 inches of water differential) limit application to fairly steady rates of flow (low rangeability). These devices are very inexpensive and may be installed so they can be removed for maintenance without stopping line flow.

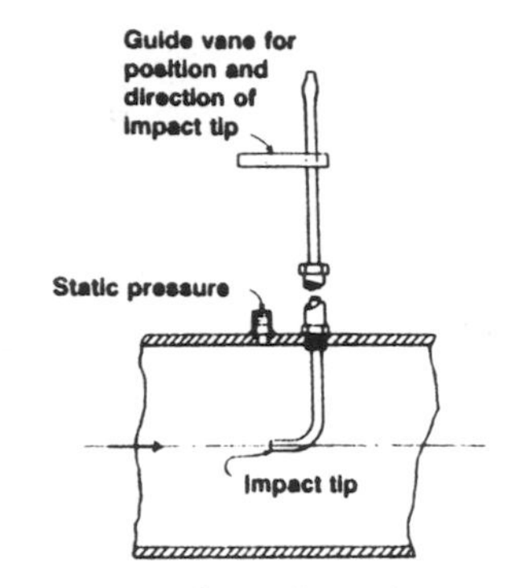

Figure 8-19 *Pitot tube or impact meter.*

Advantages of Pitot meters:

1. Low cost for installation and operation (essentially no pressure drop)
2. Standard differential readout device for all sizes

Disadvantages of Pitot meters:

1. Point velocity measured is assumed to represent full pipeline average velocity which limits accuracy unless flow profile is closely controlled
2. For normal pipeline velocities, the indicated differential is very low which makes its accuracy poor and its rangeability very limited.

Magnetic meters are useful to measure conductive liquids or slurries. Because of the material conductivity they require for operation, they are not useful in the petroleum industry for measuring hydrocarbons. They are very useful, however, for measuring such things as paper pulp slurries and black liquor which most other meters cannot measure. They are made in sizes from fractions of an inch through almost 100 inches. Since they operate on velocity, their equations to convert from flowing to base conditions are essentially the same as for turbine meters.

Meter operation is such that density and viscosity do not directly affect its operation. The meter operates bidirectionally provided upstream lengths are used on both sides of the meter to control the velocity pattern. Since the meter is full line size there is no pressure drop caused by the meter other

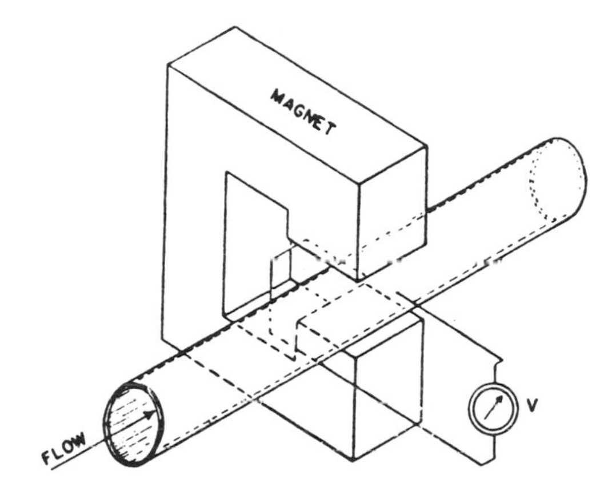

Figure 8-20 Elements of an electromagnetic flowmeter.

than normal pipe loss. The meters are fairly expensive and have a high operating cost because of the high power requirements. In the large sizes, they are quite heavy and require special considerations for installation.

Advantages of magnetic meters:

1. Performance not affected by changes in viscosities and densities
2. Full-bore opening means no head loss
3. Meters will operate bidirectionally with required upstream lengths installed on both sides of meters
4. Available with insert liners that allow use on some corrosive and erosive fluids

Disadvantages of magnetic meters:

1. Installation and operating costs relatively high because of size, weight, and electrical power costs
2. Fluids must have at least the minimum conductivity specified by manufacturer of the specific meter
3. Used for liquids or slurries but *not* gases.

Coriolis meters can be used on liquids and some gases. They directly measure weight. If the desired measure is volume, then some correction for density at fluid base conditions must be made. Some models of the meter offer both mass rate and density from one device. Since these meters react to mass, they can be used (within limits) for mixtures of liquids and gas. The manufacturer should be contacted for recommendation on a meter's limits with mixtures since such an application presents special problems.

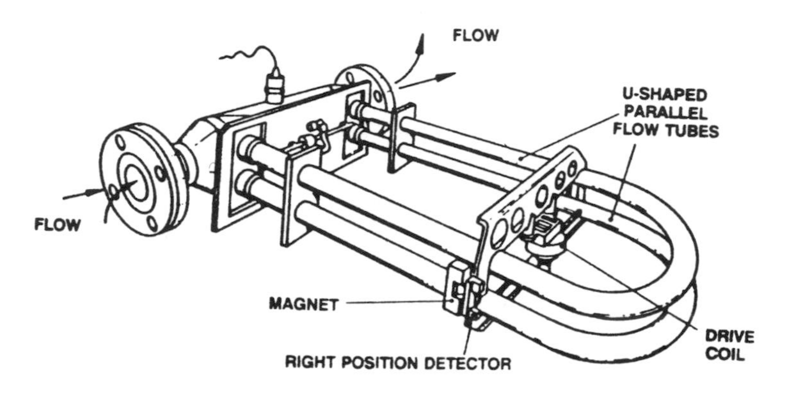

Figure 8-21 *Modern industrial Coriolis mass meter.*

The meters are most popular in sizes of 1/16 inch through 6 inches. They are offered in larger sizes, but mechanical limitations begin to become onerous in larger sizes. These meters are used for both custody transfer and control measurement. They are relatively expensive compared to other choices. However, maintenance costs are minimal provided operation is on clean fluids. Any depositions or collections of materials in the meter body will be indicated as a density error. However, the flow tends to keep the meter swept clean.

Advantages of Coriolis meters:

1. Can be used on liquids, slurries, gases, and two-phase liquids and gas flows (within set limits)
2. Units measure mass directly, an advantage when mass measurement is desired (See 2 below)
3. Can handle difficult fluids (highly varying densities or phase mixtures) where other meters cannot be used (Should check with manufacturer for suggestions and limitations in such applications)

Disadvantages of Coriolis meters:

1. Available only in 1/16 through 6 inch sizes
2. If volume measurement desired, conversion via analysis or density measurement at base conditions required
3. Special installation requirements to isolate meters from mechanical vibration
4. Pressure loss across meters may be high

Other types of meters with special uses

In addition to the more common meters covered above, other meters are available that are worthy of mention and a short description. When an application is contemplated, specific information should be obtained from suppliers of the meter.

Elbow Meters

As fluid flows around an elbow, centrifugal force makes pressure on the outside wall higher than pressure on the inside wall. This pressure difference is proportional to flow, and its coefficient can be estimated from knowledge of elbow dimensions. For more accurate measurement, an elbow (with at least 10 diameters of straight pipe upstream — straightening vanes are recommended to stabilize swirling flow — and 5 diameters downstream) should be flow calibrated. If welds on an inlet elbow and pressure taps are carefully made, elbows will calibrate with a very stable calibration curve.

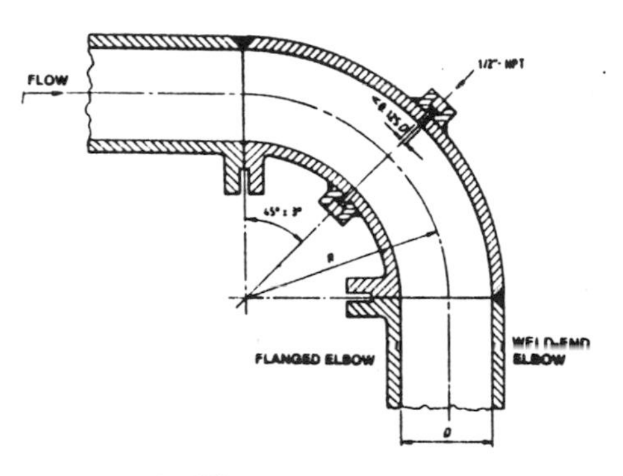

Figure 8-22 Elbow meter.

However, these units are more often used for flow control (high repeatability) rather than for accurate flow measurement. Piping systems already have elbows present, and their use as a meter adds no pressure drop not already present. But normal pipeline velocities do not generate high differentials (normal maximum about 9 inches), and this limits accuracy and severely limits rangeability.

Elbows are normally used for control with stable flow rates where relative rather than absolute values of flow are required.

Swirl Meters

A fixed-geometry helix blade in this type of meter imparts a swirl upstream of a Venturi-shaped throat that increases the stream velocity, followed by a deceleration in an expanding cone. This action generates a precessing vortex (swirl) whose frequency is a function of flow velocity through the meter. A sensor picks up the temperature change in the swirl or variation in pressure caused by the swirl. The meter is normally used only on gas flows because of its high head loss. Also, the complexity of mechanical construction involved requires calibration for each meter for best accuracy.

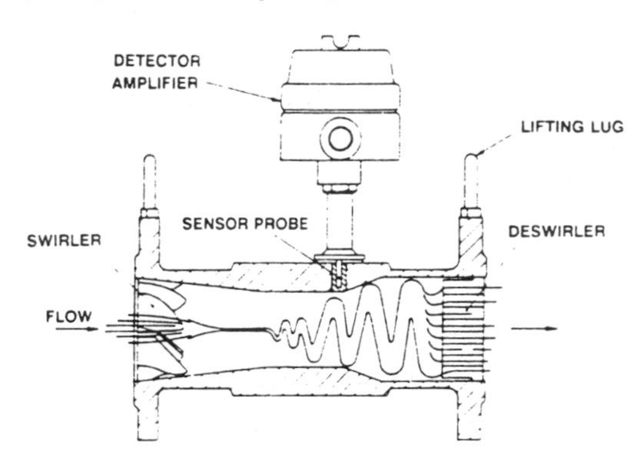

Figure 8-23 Swirlmeter.

Insertion Meters

Three types of insertion meters are based on previously covered meter types: turbine, vortex, and magnetic. Use of insertion meters is to sample a point velocity representative of the average velocity of a full stream. This limits their absolute accuracy to the validity with which the velocity sample point is located. However, their repeatability may be sufficient for some uses as a control device. Because of the small amount of blockage, insert meters cause small-to-no pressure drops compared to drops of equivalent full-bore meters. (Refer to previous material in this chapter for application suggestions and limits.)

Special-Application Meters

Certain flow meters have specific applications but are not considered for general use. The exclusion may be for cost or because the meter is newly developed or limited by its design/operation. This does not mean that at some future time there may not be further developments to expand a meter's use. Such meter types include thermal, tracer, laser doppler, nuclear magnetic resonance (NMR), and sonic nozzles.

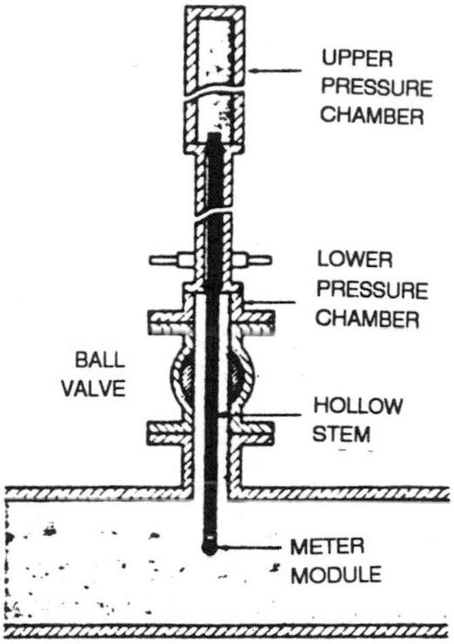

Figure 8-24 Insertion-type turbine meter.

Thermal meters have been used for some research applications. However, developments are now beginning to make them attractive for some commercial flow jobs. These meters follow two basic operational principles:

1. A body is heated by known heat input, and the body is cooled by the flowing stream; this temperature change is proportional to mass flow rate.

2. A heat source adds heat to the stream, and downstream sensors measure the temperature rise which is proportional to mass flow rate.

The hot-wire anemometer is used to define flow velocity in a clean gas stream flow as the fluid cools the element. The element is operated at constant current, and its resistance is kept constant so voltage variations measured relate to velocity. In either case, sensitivity of the element to fouling requires frequent cleaning even in clean streams; this limits the practical application of this type of meter in commercial measurement and makes it useful mainly in research.

Another way to use thermal energy for measurement is to install two thermistors in a flow stream, one in the flowing stream and the second in a side pocket out of the flowing stream. The temperatures of both probes are kept constant, and a measure of the difference of power supplied is proportional to the flow rate.

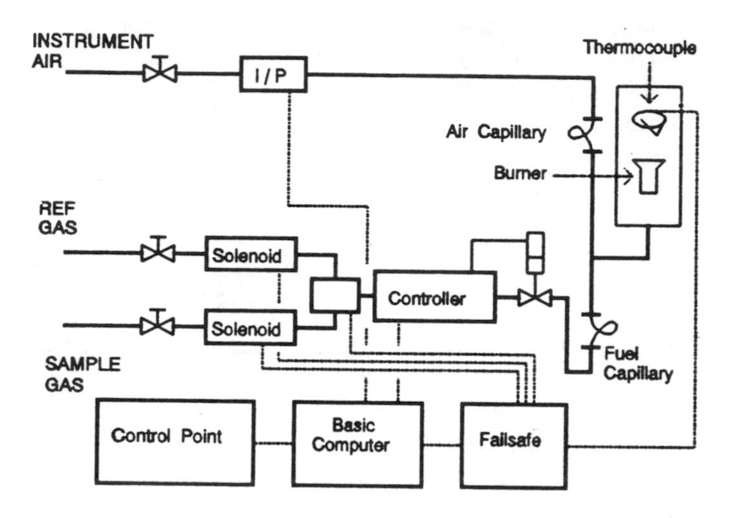

Figure 8-25 Thomas thermal flowmeter.

A similar meter measures the temperature before and after a heat source with both probes in the flowing stream.

All of these meters are so sensitive to problems of dirty streams that they find application mainly in determining low velocity mass flow rates.

Tracer meters have been around for many years. The first units injected a foreign substance into a flowing stream and then picked up its presence at one of two detectors downstream of the injection. One of the first meters injected salt into the water. The details of injection, detection, dispersion, flow profile, distance between probes and marker types have given rise to many different meters and make-ups. They are used for spot checks of velocity and, with an area factor, allow volume calculation.

A marker must have some characteristic that sets it apart from the flowing stream — such as the salt in fresh water, radiation, heat, and dye properties. The tracer should be approximately the same density as the flowing stream. It should mix well and travel at the same speed as the carrier. It should be readily available at low cost, chemically inert, and not naturally present in the stream. It should be detectable by some standard analysis technique such as conductivity, color, radioactivity, chromatographic analysis, flame ionization or photometry and heat sensing.

Most tracers operate intermittently by manual injection, but some tracer meters measure flow continually with automatic injection based on time or marker detection. Injection and detection points are based on distance and dispersion of the sample matched against detector sensitivity. The most accurate measurement with these devices requires calibration in-place, but they can be used with less accuracy for lines already in place by adding the injector and detector(s). They are quite often used in control measurement to set flow rates where the absolute value of flow is not critical.

Laser Doppler meters are similar in concept to doppler ultrasonic meters but more expensive and difficult to set up. For these reasons, they are more often used in research facilities for flow measurement. They require transparent pipe and flowing fluids which allow light to penetrate the flow stream.

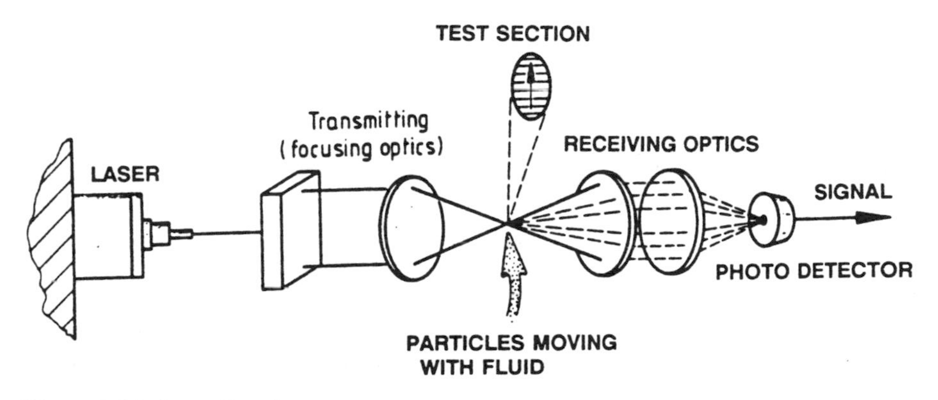

Figure 8-26 *Laser Doppler anemometry.*

Two light beams are focused on a particular area where flow velocity is to be measured. Any light-sensitive particles that pass this point scatter the light which is measured by a photo detector. The particle velocity causes a doppler shift that produces a signal in the detector proportional to flow velocity at that point. Beams are moved across the flow to enough points to establish an average flow velocity.

These devices have definite use in studying flow profiles and patterns but have little use in converting a flow pattern to flow volume.

Sonic Nozzles are special shaped nozzles used for calibrating gas flow devices. A sonic nozzle is used mostly as a test device because its high

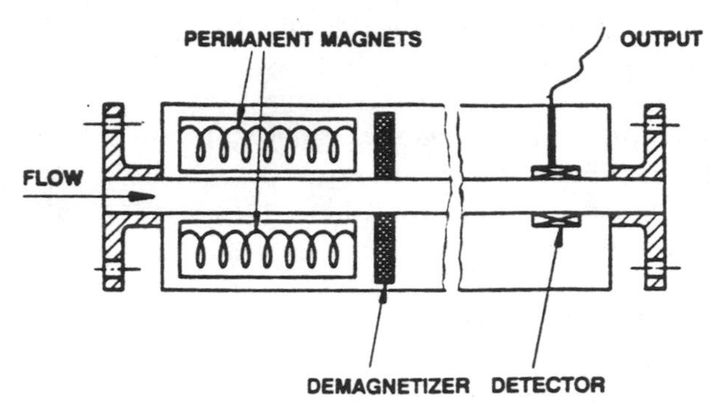

Figure 8-27 Schematic of a nuclear magnetic resonance meter.

pressure drop (10 to 15% of inlet pressure) is too costly to absorb in most operations. It is quite accurate, providing thermodynamic properties of the flowing gas are known accurately. The most common use is as a calibration device for natural gas meters such as positive displacement or turbine meters used at over 35 psia. Details on meter construction, calculation procedures and use are available from meter manufacturers. The nozzles measure only one flow rate for a given static pressure and therefore are not used for normal flow measurement.

Nuclear Magnetic Resonance (NMR) meters mark the nuclei of hydrogen or fluoride in a flowing fluid. The fluid then enters a detector section when the magnetized nuclei relax between two detectors. This measurement produces a frequency proportional to fluid velocity within the detector section whose length and volume are known.

These devices, in development infancy, are very expensive. However, their unique characteristics allow them to cover a wide range of difficult-to-measure flows (i.e., slurries, non-Newtonian fluids, and emulsions). Industrial applications are limited with most uses in development and research work.

References

1. *American Gas Association, AGA-3 "Orifice Metering of Natural Gas and other Hydrcarbon Fluids. 1515 Wilson Boulevard, Arlington VA 22209.*

2. *American Society of Mechanical Engineers Fluid Meters, Their Theory and Application. New York: ASME, Sixth edition, 1971.*

3. *American Petroleum Institute Manual of Petroleum Measurement Standards. Washington: API, Chapter 5 "Metering."*

4. *American Gas Association Measurement of Fuel Gas By Turbine Meter. Arlington VA: AGA, 1981.*

5. *ANSI/API 2530 (AGA-3) Orifice Metering of Natural Gas and Other Hydrocarbon Fluids. Washington DC: ANSI/API, 1985.*

6. *American National Standards Institute/American Society of Mechanical Engineers Performance test codes 6. New York: ANSI/ASME, "Steam Turbines", 1982.*

7. *American Society of Mechanical Engineers Differential Producers used for the Measurement of Fluid Flow in Pipes by Orifice, Nozzle and Venturi. New York: ASME, "MFC-3M, 1982.*

8. *International Standard Organization ISO 5167, Measurement of Fluid Flow By Means of Pressure Differential Devices. Geneva, Switzerland: Part 1 "Orifice Plates, Nozzles and Venturi Tubes Inserted in Circular Cross-Section Conduits Running Full," 1991.*

9. *International Standards Organization ISO 9951, The Measurement of Gas by Turbine Meters. Geneva, Switzerland: ISO, 1991*

9
READOUTS AND RELATED DEVICES

Secondary systems are a part of any measurement installation for reading primary element signals and the variables necessary to correct flow from flowing to base conditions. These elements fall into three main categories: mechanical, pneumatic and electronic. All have applications in flow measurement. The choice can depend on a number of parameters not the least of which may be personal preference based on a given industry's experience. The fastest growing segment is the electronic systems designed to take advantage of the rapidly growing availability of computers.

Electronics

Several alterations have taken place in the move to electronics. First was the glamour of being "up to date." Using the new electronics at that point often created about as many problems as were solved. The common reaction was "electronic equipment doesn't work". Designers then came up with new developments and improved capabilities, and users began a more studied evaluation of true needs and uses. These evaluations defined the most useful areas in which to apply the devices.

The present generation of transducers and computers are well received by users, more and more of whom are converting to electronics each year. There are several reasons for this user acceptance. Service provided is equal to or better than mechanical and pneumatic types. Many additional capabilities are possible (i.e., smart transducers, etc.). Minimum maintenance is required. Trained personnel are available to install and maintain the units, and power requirements have been reduced to the point that auxiliary power sources such as solar-charged batteries may be used for remote locations.

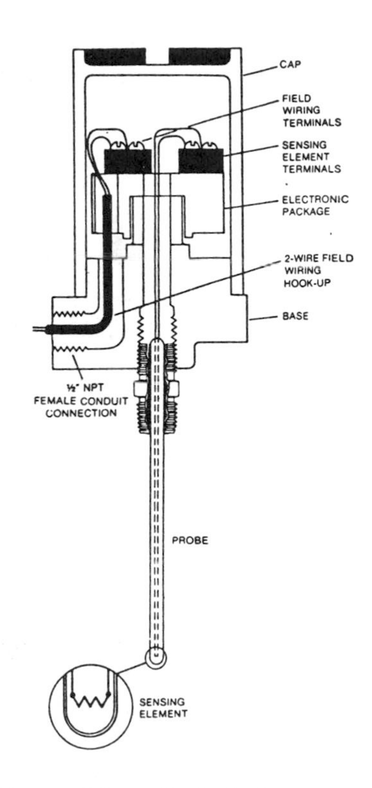

Figure 9-1 *Typical temperature transducer. Devices such as this, along with computers and other equipment, are vital parts of a metering system.*

"Simple" computers may simply calculate a flow rate and totalize flow for a meter. Or computers can be operating centers for measurement, control and communications in complex multimeter systems. Computers can develop the complete volume calculation and print appropriate hard copy or feed a central control or computer center with the complete accounting procedures. They can provide real-time operation and control information for metering systems.

Their primary limitation is cost justification versus alternative ways to achieve desired measurement over a given service life. Systems for individual meters may cost only several hundred dollars, whereas larger systems can run over half a million dollars. Cost effectiveness usually boils down, not to accuracy considerations, but rather to efficiency of solving equations and the true need for speed.

The manner in which equations are broken down in standards and references indicates that their values are independent and are individually addressed. This is an inherited interpretation based on capabilities of past

Figure 9-2 *Typical orifice meter with electronics for flow control measurement.*

equipment. As long as conditions are relatively stable, the use of time averages does not introduce significant errors. For varying flow, however, the basic equation requires the variables to be interpreted on a continuing basis with the readout system's frequency faster than the flow system change. Such changes can be quite rapid, and the frequency response requirement demands use of an electronic system for flow measurement. A system that doesn't change much over hours of operation, on the other hand, can provide measurement without requiring continual integration from electronics related equipment.

Related Devices

Most related devices have an accuracy in the general range of ± 0.5% to ±1% *of full scale.* This makes it important to choose transducers with the right ranges; measurements should be in the upper two thirds of device ranges. The higher the differential the better, provided flow conditions do not exceed the differential device's range. Many users, not recognizing the effects auxiliary instrumentation can have on the accuracy obtained by a given meter, compromise flow accuracy accordingly. Overall accuracy obtained includes the inaccuracies of each of the auxiliary devices and how it is used, as well as the primary meter accuracy. Sometimes the auxiliary

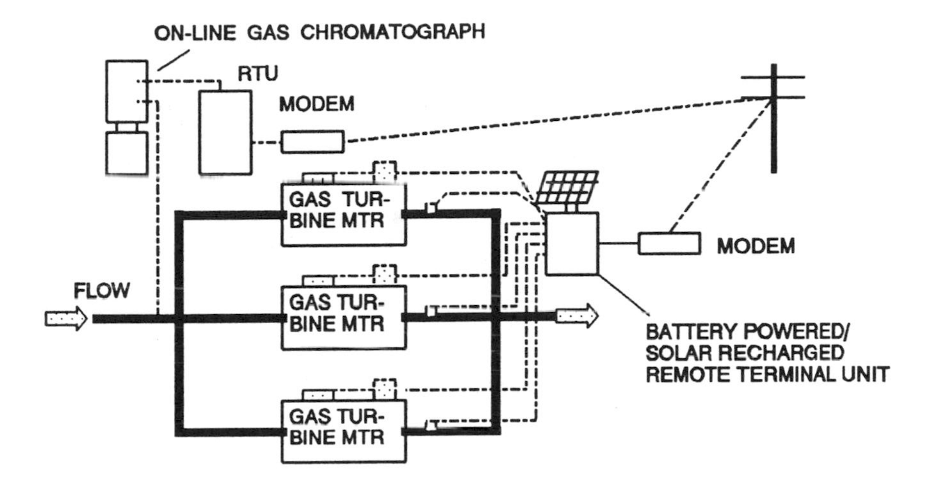

Figure 9-3 Typical multiunit flow computer system with turbine meters.

devices may control results more than the basic meter. For example, with gases (particularly near their critical points), a change of one pound can represent several percent in flow. This says two things: 1) it would be better to measure at some other location where conditions are farther away from critical points, or 2) the accuracy required for pressure measurement must be increased severalfold to maintain the same limits for the corrected flow measurement.

Pressure measurement for liquid flow is straightforward. Liquids generally are less sensitive to their pressure measurement. However, in areas near the liquid critical point, density changes significantly and affects flow measurement accuracy.

Location of the pressure tap for a meter is based on the meter calibration in the same manner as one of the differential pressure taps for an orifice meter with gas flow. The tap into the line should be at a point specified so that flow past the tap creates no undue turbulence which can affect the reading.

The ideal point to measure pressure (at the point of velocity or volume determination) is usually not possible — or at least not practical — with most meters. When such mechanical problems make it impossible to install a pressure-measurement tap at the proper point of flow volume or velocity determination, then corrections may be required to account for the difference between the theoretical point and the actual point. Sometimes the difference is simply ignored if the difference does not affect density calculation "seriously." The orifice equation for gas has an expansion factor in it to make the required correction. The difference can also be accounted for as part of basic meter-system calibration.

Figure 9-4 Pressure transmitter.

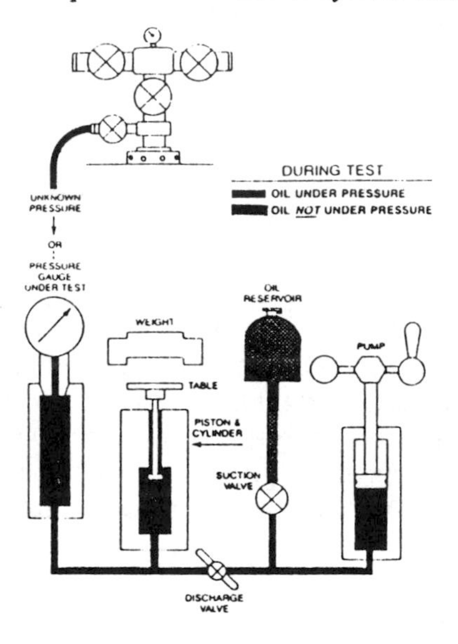

Figure 9-5 Schematic of a dead-weight tester, a standard for pressure measurement.

Pressure transducers must be calibrated on a routine basis to maintain accurate flow measurement. The standard most commonly used is dead-weight testers or precision gauges for higher pressures, manometers for lower pressures.

Temperature transducers present problems similar to pressure transducers. Since volume flow normally is not at base conditions, a measure of the flowing temperature is required to correct for the difference. The accurate measure of temperature is more difficult than it appears since the transducers normally require insertion into the flowing stream and thus disturb smooth flow which consequently disturbs meter operation. Therefore, temperature is normally measured at some point downstream of the meter after making sure that the temperature will be essentially the same as the temperature in the meter.

The effect of ambient conditions on the readout equipment is also important. Radiation from the sun and conduction from pipeline heat can effect temperature readings by changing the temperature of thermowells and/or affecting transducer mechanics. For utmost accuracy, the instrument environment should be controlled with heating and cooling, shading, or insulation, depending on the required flow accuracy and the effect that temperature has on the fluid properties and flow measurement. Fluids near critical points are prime candidates for this treatment, whereas other less critical fluids generally require no unusual treatment. Smart transducers have helped minimize some of these ambient effects at less overall cost than some of the other treatments listed.

Differential pressure is the most important variable for differential head meters, since most errors in flow measurement with differential meters come from this measurement. These errors are so critical because differential pressure is the major factor in calculating flow. The maximum differential used with these meters is in the range of several hundred inches of water (i.e., less than a 10 psi drop). Quite often, static pressures may run hundreds of pounds, so the measuring device has a static pressure load on it of about 1000 psia — yet it is trying to measure a difference in pressure of 1 percent or less of its static range. This requires the differential device to be rugged enough to withstand the high pressure requirement, yet sensitive enough to measure a very small pressure difference.

To minimize the problems (consistent with the required flow range), operate with the differential at higher values — provided the strength limit of the pressure-drop creator is not exceeded. Because of the cost of lost pressure, differential pressure devices commonly used are in the 100 to 200 inches of water range. The devices used include manometers (used at low static pressures), diaphragm bellows, and mercury filled meters. These devices must be calibrated against manometers or dead weights. They are usually tested and calibrated at atmospheric pressure and then zeroed at line pressures. Some test devices that operate at line pressures are available, but their use is often restricted to laboratory work rather than field calibration.

Maintenance of differential devices consists of periodic calibration and, if necessary, replacement of driving mechanisms. In dirty service, periodic cleaning may be necessary or use of seal pots or isolating diaphragms may be required to prevent contamination.

Where wide ranges of flows are expected, multiple transducers can be used on a single meter to expand its range. For example, a more accurate low differential device, such as a 20-inch water unit, can be manifolded into the same meter as a 200-inch unit. This combination expands the range of

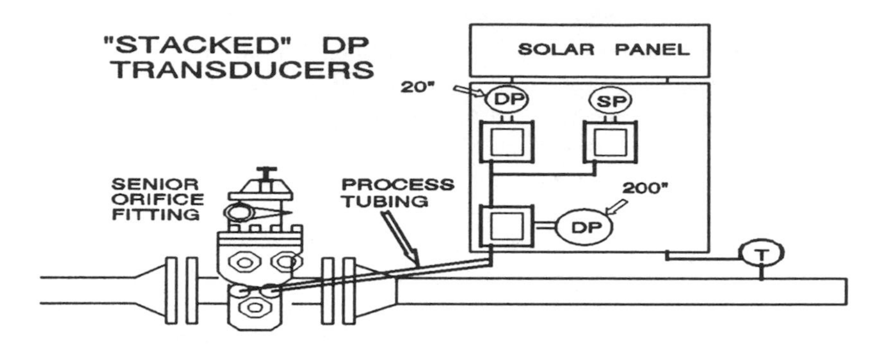

Figure 9-6 Stacked transducers may be required to extend a measurement range.

flow from 3-to-1 (on a 100-inch unit) to approximately 10-to-1. If ranges beyond this are required, then a second or third meter with proper valving can be used with meters switched in and out as the flow varies. Combinations of this sort allow an almost infinite flow range to be handled.

Once again, for the most precise flow measurement, the use of smart differential devices is an investment of value to minimize the effect of ambient conditions on the device.

Specific Gravity. Reducing the fluid from flowing conditions to base conditions requires identification of fluid composition. The most useful measurement for this is specific gravity or relative density of the fluid. Correlations in the petroleum industry are based on these measurements, and data for other mixtures are expressed in these terms. For pure products, the need for specific gravity reduces itself to the relationship of specific gravity at flowing conditions to base conditions which corrects for the effect of pressure and temperature on the pure product. Quite often this is available in a formula expressing the effects of pressure and/or temperature, and specific gravity correction can be made with suitable measurements.

The several definitions of specific gravity used in the flow measurement business are important to understand. For natural gas, the definitions in AGA-3[1] are the weight per unit volume of gas compared to the weight per unit volume of air at the same conditions of pressure and temperature. This definition of "real specific gravity" ignores the corrections for compressibility when these weighings are made at atmospheric pressures since they are relatively small. However, this yields a specific gravity different by a small

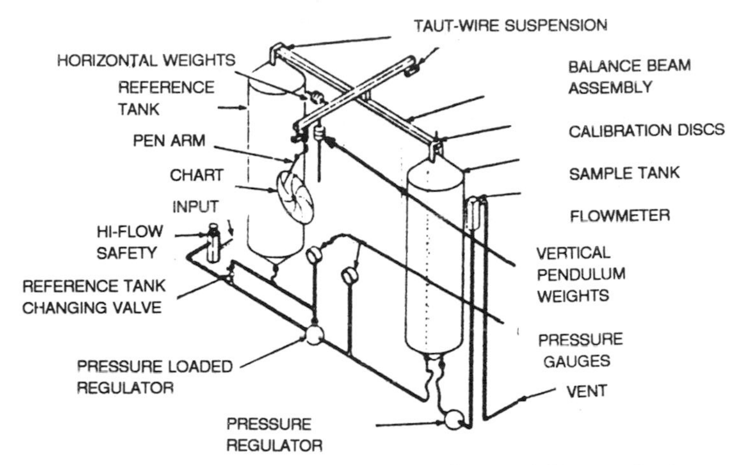

Figure 9-7 *Recording gravitometer using indirect weighing method.*

amount from the ratio of molecular weights (which is equal to the ideal specific gravity). In non-natural-gas measurements, these definitions are not made. And the normal definition used for specific gravity is the ideal AGA-3 definition (i.e., the ratio of molecular weights).

In liquid measurement with the English system of units in the U.S., definition of specific gravity is different in that the weight per unit volume of the liquid is compared to the weight per unit volume of *water at 60°F.* Water at 60°F has a definitive weight set by the International Steam Tables, so that a liquid specific gravity is directly convertible to density by multiplying the weight of water at 60°F times the specific gravity of the fluid.

$$\delta_f = (SG)(W_w) \qquad \textbf{(30)}$$

where δ_f = flowing density (or specific weight)
SG = specific gravity
W_w = weight of water at 60°F

This calculation is not possible with natural gas since there is no specification for the base air conditions, and hence no specific weight may be assigned.

When specific gravity alone does not sufficiently represent composition for flow calculations, then an analysis is required. This can happen when variable components may make up a sample with the same specific gravity. Natural gases and mixed petroleum liquids exemplify the problem.

Sometimes there is a need to know the constituent make-up for pricing information if each component has a separate value. Corrections may be made for nonhydrocarbon constituents in the streams.

Sampling

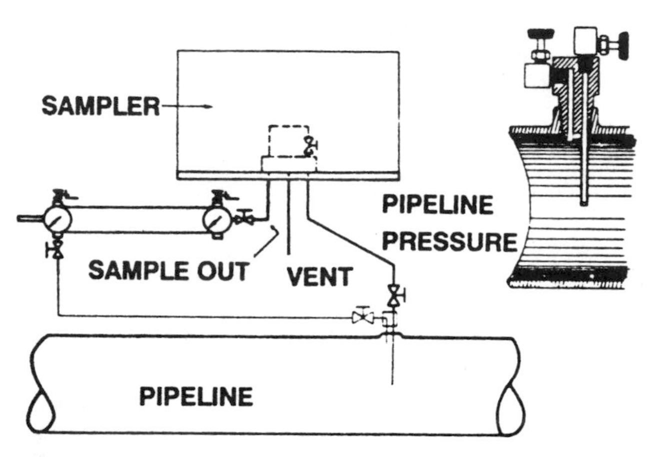

Figure 9-8 *Typical sampling system.*

The major concern in running an analysis is to get a representative sample of the pipeline fluid into a sample container and then into an analysis instrument. Sampling is a science unto itself, and great care must be taken to get a representative portion of the flowing stream for testing. Once a sample is in a container, getting it back out of the container can also create error. Fluids difficult to sample include light hydrocarbon liquids (gases at atmospheric pressure and ambient temperature), saturated gases (water or hydrocarbons), gases containing hydrogen sulfide, condensing gases or vaporizing liquids, crude oils containing water, and emulsions. In these cases, special procedures and equipment are required even to attempt sampling. But even with these considerations recognized, getting good samples requires perseverance and luck. In the most sensitive cases, direct sampling into the analysis equipment is required. It is no field for an amateur to enter. For example, values involved in the petroleum industry affect the exchange of money — and purchasers don't care to pay for crude oil but get water measured as crude oil with improper sampling.

Analysis allows calculation of parameters important to flow measurement to be calculated such as: specific gravity, heating value, compressibility factor, inert content and density. Calculations are based on mixture laws and are accurate at base conditions, but conversion to flowing conditions is not easy and can, in certain circumstances, introduce errors where the mixture laws break down because of shrinkage (such as mixtures of light hydrocarbon liquids).

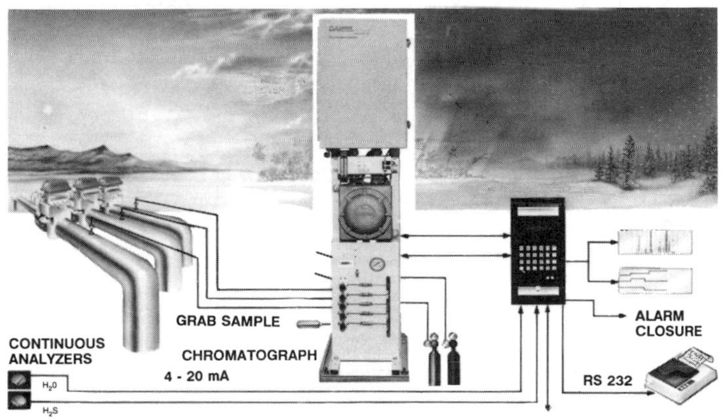

Figure 9-9 Gas chromatograph system with electronic computer/controller.

The most common analysis instrument is the chromatograph. Based on standardized samples, chromatographs can be calibrated to cover wide ranges of fluids. Easily maintained, their calibration can be checked with a standard sample wherever desired. The units come in models that can be operated continuously or intermittently when a sample is available. Most are permanently installed, but portable units can be used as line monitors at strategic locations until a problem arises elsewhere; then the portable unit can be taken to the problem site for on-the-spot analysis.

Calorimetry Where heating value is needed, a calorimeter can be installed to continuously monitor a stream. Or, samples can be taken at meter locations and individual samples tested at a centrally located calorimeter. After all inputs have entered a pipeline, a single unit is often used to

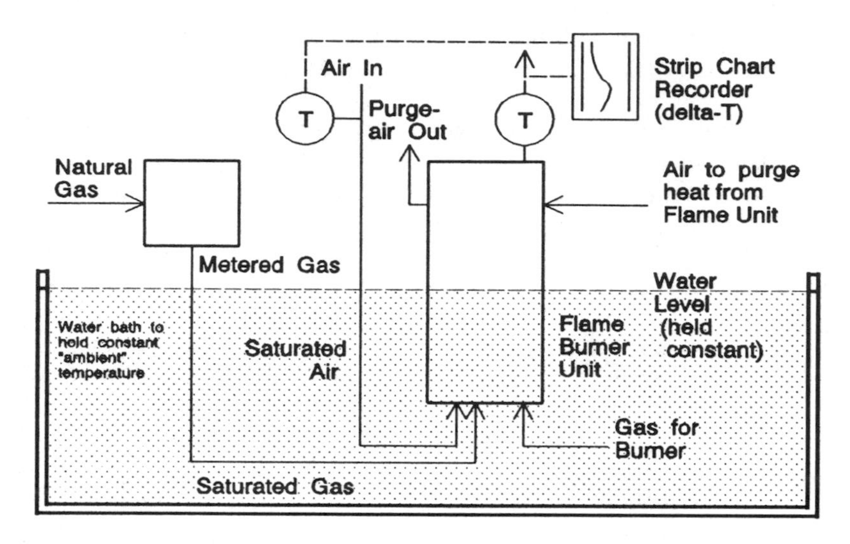

Figure 9-10 Schematic of principal elements of Thomas type calorimeter.

determine the heating value at all downstream locations. The decision as to which method to use depends on product value (quantity) and the contractual requirement for corrections (i.e. some require correction to total heat value, others only require that a minimum heat value be maintained).

Reference

1. American Gas Association Orifice Metering of Natural Gas and Other Hydrocarbon Fluids. Arlington VA: AGA, 1985

10
PROVING SYSTEMS

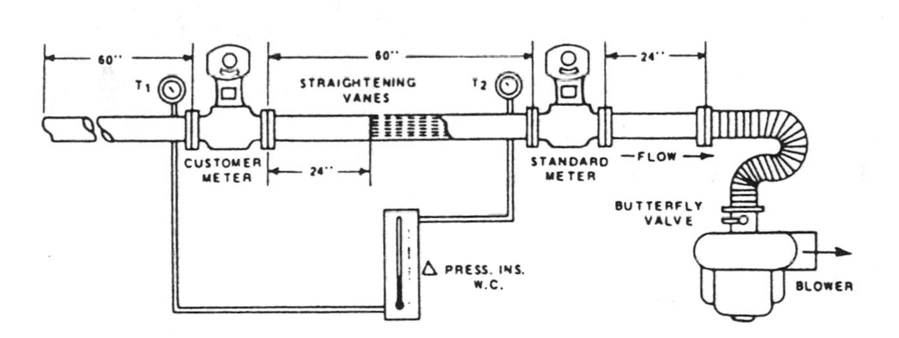

Figure 10-1 Low-pressure proving setup using blower to create gas flow.

The necessity of proving a meter depends on the value of accurate measurement for the product being handled. Large volumes and/or high-value products are the prime candidates for using provers. Oil industry measurement of crude oil and refined products are examples of where meters typically involve proving systems[1]. The proving systems are considered part of the cost of the meter stations and are permanently installed at large facilities. When product value is lower, provers are usually portable (used within a limited geographical area); as product value drops further, proving frequency is reduced, and for the lowest value products proving is not done at all.

In other industries, proving in place is seldom done; metering is assumed correct until a process goes out of control or a meter breaks down and requires repair or replacement. For meters such as the orifice type, calibration is accepted as correct as long as mechanical requirements of the meter's specifications are met. Some meters are tested by calibration of readout units only, with no test or inspection of the primary device.

In summary, testing can be a very expensive and time consuming proving procedure, or it can be as simple as a physical examination during a walk by. Obviously, the ability of these tests to prove a flow meter's accuracy varies from the best that can be done to a "test" that really has nothing to do with flow accuracy. One of the easiest meter proving methods is to check the operating meter against a master meter that has a pedigree of accuracy. Any differences indicated can be calibrated into the operating meter by use of a meter factor applied manually or through a meter adjustment. Meter factor in this case is defined as:

Figure 10-2 Master meter prover.

$$\text{Meter Factor} = \frac{\text{True Volume}}{\text{Indicated or Meter Volume}} \quad \textbf{(31)}$$

which "divides out" the meter indication and "multiplies in" the corrected volume.

Since most meters are not totally linear, tests should be run over the meter's operating range and the meter factor entered as an average factor over the range. Computers can apply a factor which varies with flow rate. The correction complexity required depends on the magnitude of meter non-linearity and the measurement's accuracy requirements.

Liquid Provers

For products that have no vapor pressure at flowing conditions, an open tank prover may be used as a standard. Calibrated cans (seraphins) are calibrated and stamped by standardizing groups with the volume they contain or deliver. Flow from a meter is diverted from normal delivery into the can until it is full. Readings of the operating meter are taken at the start and upon completing the filling. This procedure can be automated by the use of solenoid valves in the fill and bypass system.

At one time, similar systems for test fluids with vapor pressures were used with closed containers. But the cumbersome cans required have been replaced with the more convenient pipe provers. Pipe provers are available in several configurations such as standard and small volume, u-shaped, straight or folded, ball or piston displacer, unidirectional and bidirectional. The choice depends on parameters of the job to be done.

Figure 10-3 *Liquid prover system.*

The bidirectional prover requires a displacer round trip to complete one prover run. It can be made U-shaped, folded or straight shape depending on space requirements.

The standard prover (U-shaped bidirectional) is the most common and uses an inflated ball displacer. Regardless of construction and operation details, all provers perform the same function. Flow is passed through an operating meter into the prover. When temperature and pressure have stabilized, the displacer is launched. Since this creates a temporary slow-down in flow until the displacer gets up to speed, some prerun length must be allowed before displacement of the accurately measured volume begins. At a point after flow rate stabilization, a switch indicates entry of the displacer into the calibrated section, and the meter pulses are sent to a proving counter or circuit.

Flow continues until a sufficient number of pulses (typically 10,000) has been generated by the operating meter. An exit switch then indicates that the calibration volume has been achieved, and pulses to the proving counter

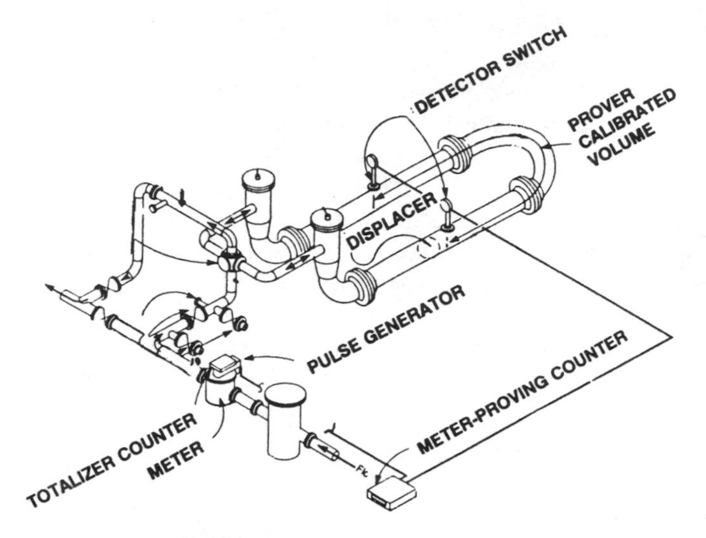

Figure 10-4 *Typical bidirectional U-type sphere prover system.*

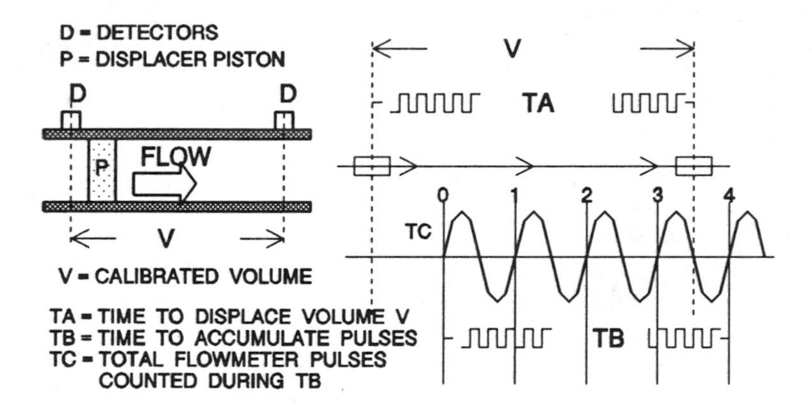

Figure 10-5 *Schematic of pulse interpolation.*

are interrupted. Pulses generated by the operating meter are thus "gated" to the proving counter, without stopping the same pulses from going to the billing meter's counter. This displacer passage and collection of pulses is repeated a number of times (set by individual company policy but typically four or five times) while recording the stabilized fluid pressure and temperature. Calculations convert the temperature and pressure to the same or base conditions for the meter and the prover. When volumes are compared, the ratio of the prover to meter volume is the meter factor for this flow rate.

Various provers have distinguishing characteristics.

The small volume prover has a precise pickup system that allows less tolerance in the switch location for the displacer. The timing system requires fewer meter pulses to generate the necessary pulse rate to achieve 10,000 pulses between switches. To minimize the prerun and flow interruption when the displacer is launched, the displacer is driven externally rather than taking its energy from the flowing stream. This allows stabilization to be reached rapidly. This system allows proving with less volume displaced, hence the name "small volume" prover.

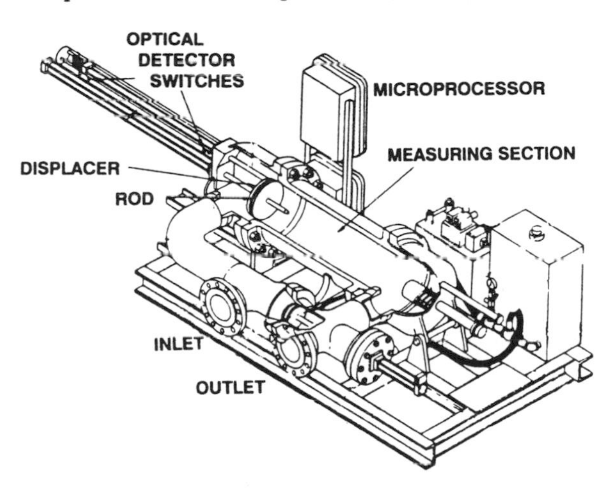

Figure 10-6 Typical small-volume prover.

The unidirectional prover, built to send the displacer in only one direction, requires a volume large enough to produce the 10,000 meter-generated pulses.

The piston displacer prover uses a straight barrel design since the piston can't go around a corner. It is used when the fluids, because of composition or temperature, make "standard" displacers unusable. The piston can use seals that will operate at temperature extremes and on most corrosive or reactive fluids. Since the seals operate on a smoothly machined and coated surface, the fluid stream should contain no erosive particles. Whatever the job to be done, a prover can probably be made to meet the requirements. As product value has climbed through the years and prover costs have dropped, many industries who in the past didn't use these devices are now using them to improve their measurement.

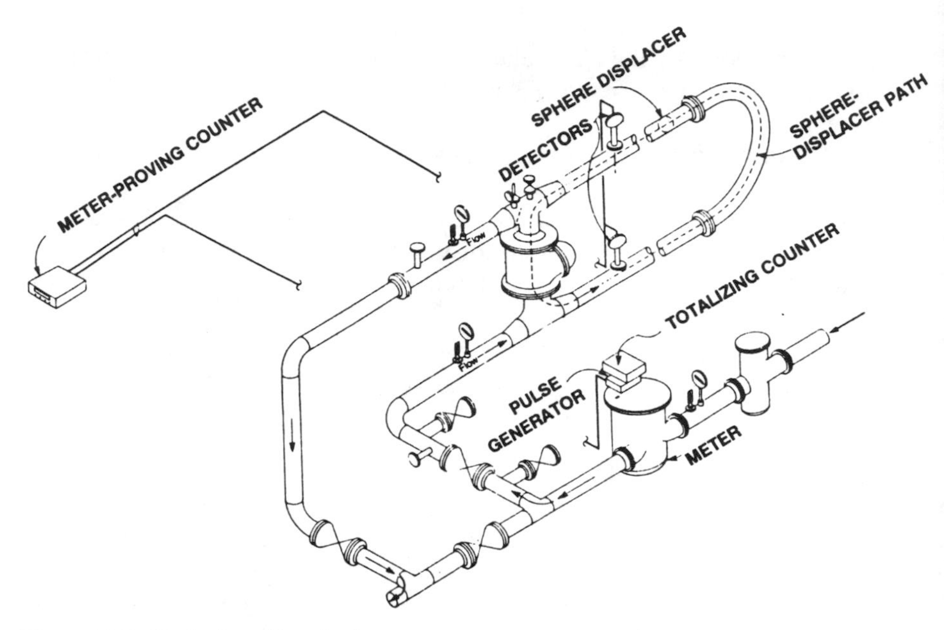

Figure 10-7 Typical unidirectinal return-type prover system.

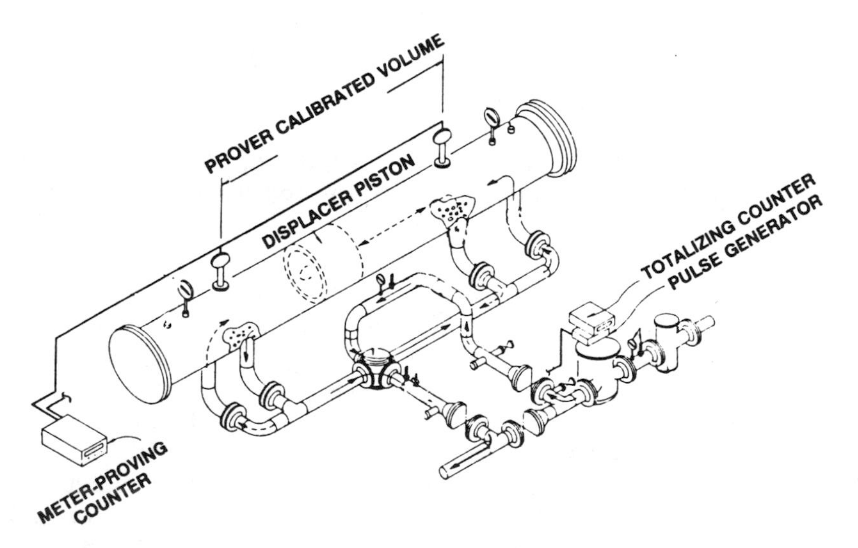

Figure 10-8 Typical bidirectional straight-type piston prover system.

Gas Provers

In the past, gas meters have not been proved like liquid meters. Proving an orifice meter has meant making sure the meter's physical condition is maintained. In looking at ways to lower tolerances on gas meters of all kinds, particularly to reconfirm a meter or settle a concern over an individual meter's accuracy when physical inspection is not sufficient (such as with a P.D. or turbine meter), actual throughput testing with a prover is becoming more prevalent. Some operators are also beginning to use them with orifice meters. These provers may be master meters, critical flow provers, pipe provers, or a centralized proving facility where meters can be taken for accuracy confirmation.

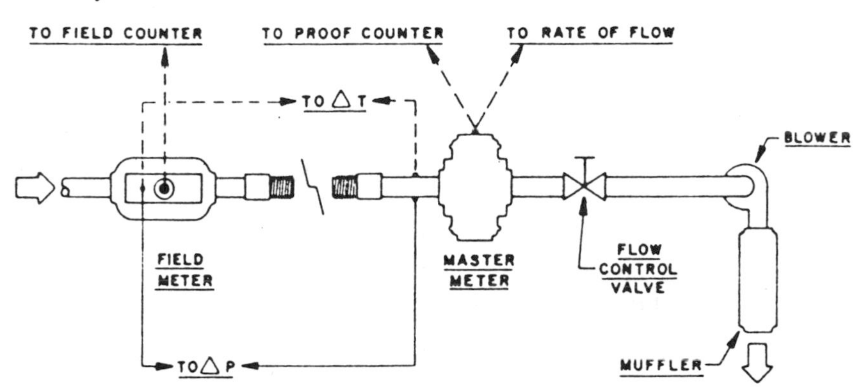

Figure 10-9 Schematic of vacuum (air) tester.

Master Meters are meters, whose basic calibration has been certified, that can be placed in series with an operating meter for a comparative test. They can be made up in special test units with a computer to control the equipment, collect the data and calculate a meter factor. For small, low pressure meter testing, they can be equipped with a blower. The meter is taken out of service, depressured, and piped in series with the proving unit downstream. The blower then pulls air through the operating meter and the standard meter to obtain proofs at a series of flow rates. A meter factor curve, plotted from these tests, allows an average factor to be obtained.

At larger volume stations with higher pressures, a master meter is typically permanently piped in series with the operating meter. A computer again controls system operation, collects the data and calculates meter factors.

Periodically, the master meter is returned to a standards lab or to the manufacturer for recertification.

These various systems have all been successfully used to improve measurement accuracies.

Critical Flow Provers One of the oldest testing devices for gas meters is the critical flow prover. These meters are installed in series with an operating meter that has been bypassed. Gas at operating pressure is passed through the meter and then the critical flow prover is normally vented to the atmosphere. If there is a nearby gas pipeline with pressure lower than the operating line by at least 15%, the gas can be passed into the second line without interfering with the test and the gas is not lost.

Several differently sized critical flow nozzles can be installed in sequence at a test holder, or a variable pressure used for testing over a range of flow rates with a single nozzle . Meter factors thus determined are used to correct readings of the operating meters.

Thermodynamic properties of the flowing gas must be known to calculate the flow at critical conditions. Usefulness of the method breaks down if the gas is near its critical temperature and pressure where correcting factors are not adequately known. Likewise, if the gas has condensed liquids or if liquids condense in the nozzle, the critical flow device cannot be used.

Critical flow nozzles are used to reconfirm meters used in natural gas pipeline and distribution systems (such as positive displacement and turbine meters).

Pipe Provers are similar to the small-volume provers used in liquid meter testing; they have been developed and are beginning to be used to determine gas meter accuracies (including orifice meters). The prover is put in series with an operating meter, and volume passed through the meter and prover with prover volume compared to the indicated meter volume so a meter factor can be determined. As the desire to reduce tolerances in measurement continues, this proving device is receiving additional attention and evaluation. Work to date indicates that, with careful testing, evaluation of a meter's operating accuracy can be obtained and a meter factor determined.

Central Test Facility Where a sufficient number of meters are in service to justify a significant test program, some companies employ a system of trading out meters (or meter internals) and bringing them to a central facility for testing on a periodic basis such as once a year. For example, offshore meters are often taken to an onshore facility for re-certification.

The centralized location where a standardized test facility is set up should have good quality gas flows available. The standard may be a master meter and/or a critical flow prover.

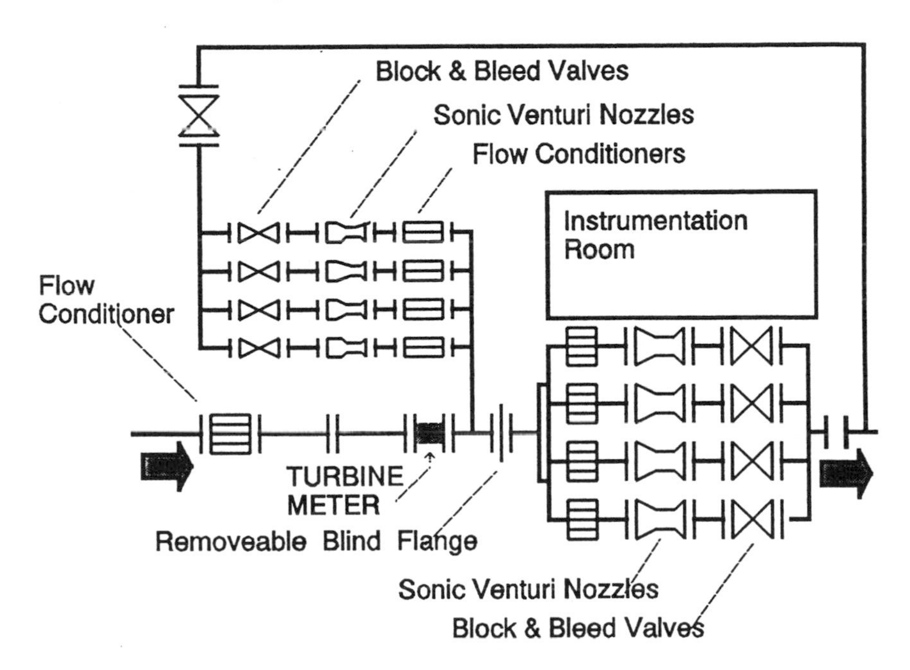

Figure 10-10 Central test facility.

Gas Proving Summary For the systems discussed above, economics must be evaluated carefully to determine justification limits for each possible method. These provings are typically done in response to governmental requirements or company policies where there is sufficient accuracy payoff to justify the expense. This normally means high volume, standard, natural gas situations or high price specialty gas systems.

Reference

1. American Petroleum Institute Manual of Petroleum Standards. Washington DC: API, Chapter 4 "Proving," 1989

INDEX

M

N

O

P